X-ray Color Imaging

Static and dynamic x-ray fluorescence for chemical element identification

Online at: https://doi.org/10.1088/978-0-7503-3215-6

IOP Series in Sensors and Sensor Systems

The IOP Series in Sensors and Sensor Systems includes books on all aspects of the science and technology of sensors and sensor systems. Spanning fundamentals, fabrication, applications and processing, the series aims to provide a library for instrument and measurement scientists, engineers and technologists in universities and industry.

The series seeks (but is not restricted to) publications in the following topics:

- Advanced materials for sensing
- Biosensors
- Chemical sensors
- Industrial applications
- Internet of Things (IoT)
- Lab-on-a-chip
- Localization and object tracking
- Manufacturing and packaging
- Mechanisms, modelling and simulations
- Microelectromechanical systems/nanoelectromechanical systems
- Micro and nanosensors
- Non-destructive testing
- Optoelectronic and photonic sensors
- Optomechanical sensors
- Physical sensors
- Remote sensors
- Sensing for health, safety and security
- Sensing principles
- Sensing systems
- Sensor arrays
- Sensor devices
- Sensor networks
- Sensor technology and applications
- Signal processing and data analysis
- Smart sensors and monitoring
- Telemetry

Authors are encouraged to take advantage of electronic publication through the use of colour, animations, video and interactive elements to enhance the reader experience.

A full list of titles published in this series can be found here: https://iopscience.iop.org/bookListInfo/iop-series-in-sensors-and-sensor-systems.

X-ray Color Imaging

Static and dynamic x-ray fluorescence for chemical element identification

Kenji Sakurai
Imaging Physics Laboratory, Ibaraki Neutron Medical Research Center E301, Tokai, Ibaraki, Japan

Wenyang Zhao
RIKEN Center for Computational Science, Kobe, Hyogo, Japan

IOP Publishing, Bristol, UK

ISBN 978-0-7503-3215-6 (ebook)
ISBN 978-0-7503-3213-2 (print)
ISBN 978-0-7503-3216-3 (myPrint)
ISBN 978-0-7503-3214-9 (mobi)

DOI 10.1088/978-0-7503-3215-6

Version: 20241201

IOP ebooks

British Library Cataloguing-in-Publication Data: A catalogue record for this book is available from the British Library.

Published by IOP Publishing, wholly owned by The Institute of Physics, London

IOP Publishing, No.2 The Distillery, Glassfields, Avon Street, Bristol, BS2 0GR, UK

US Office: IOP Publishing, Inc., 190 North Independence Mall West, Suite 601, Philadelphia, PA 19106, USA

Contents

Preface

Nature is colorful. It is full of colors, not only those of visible light but also those of other wavelengths, such as x-rays. When x-rays irradiate any object, secondary x-rays, usually called x-ray fluorescence, appear. Since the wavelength of the fluorescence corresponds to the chemical element that produced it, x-ray fluorescence is the color of the chemical element. Of course, the natural world is composed of a variety of chemical elements. Nowadays, x-ray fluorescence analysis is routinely used as a tool to see the colors of chemical elements. On the other hand, in addition to the ordinary analysis of chemical elements, imaging is also important to see the complexity and inhomogeneity of nature.

This book introduces the latest x-ray imaging that can resolve the colors of chemical elements. There are two kinds of x-ray fluorescence imaging: the scanning type and the projection type. While the scanning type is an extension of conventional x-ray fluorescence analysis, the projection type is a new trend that uses 2D semiconductor detectors. This book focuses on the latter technique because it may be more attractive to the reader. This new method enables snapshot imaging and movie imaging, which have been impossible for many years. Our aim is to provide all the information necessary to easily achieve x-ray color imaging anywhere for a variety of applications.

Chapter 1, 'The colors of the chemical elements,' explains the critical importance of developing and using sophisticated scientific observation tools. Some of the historical background of x-ray physics is described. The principle of x-ray fluorescence and some practical tips for the reader are also explained. In Chapter 2, 'What does x-ray color imaging look like?' the reader will find a number of good and attractive examples of projection-type x-ray fluorescence imaging. Chapter 3, 'Methods and instruments for x-ray color imaging,' explains the physics and technical details of projection-type x-ray fluorescence imaging. Chapter 4, 'Advanced imaging,' explains the extension of projection-type x-ray imaging to obtain more visual information, such as atomic-scale structures, valence numbers, crystal structures, layered structures, and the spatial distribution of chemical elements. The method can be used as a helpful tool for modern materials research techniques such as combinatorial synthesis. Chapter 5, 'New opportunities in x-ray color imaging,' discusses some prospects for future progress and expansion. This chapter summarizes some recommendations and advice for those who wish to try x-ray color imaging.

Kenji Sakurai and Wenyang Zhao
July 2024

Acknowledgments

The authors would like to thank Professor Sadao Aoki (University of Tsukuba), Professor Atsuo Iida (KEK), and Professor Yohichi Gohshi (University of Tokyo) for their continuous encouragement and advice over the years. The book contains much x-ray data collected in Kenji Sakurai's laboratory, which was active in the field of x-ray spectroscopy and imaging from 1988 to 2020 in Tsukuba, Ibaraki, Japan. The authors would like to thank the previous members of the laboratory who contributed to the research, especially Professor Hiromi Eba (Tokyo City University) and Dr Mari Mizusawa (Comprehensive Research Organization for Science and Society). At the end of March 2020, the authors left Tsukuba after the laboratory was closed due to the retirement of Kenji Sakurai from both the National Institute for Materials Science and University of Tsukuba. The authors would like to thank everyone in the two institutes mentioned above for their kind continuous support and cooperation, especially Dr H Kitazawa, Dr Y Yoshihara, Dr T Noda, Dr T Saito, Dr M Okada, the late Dr E Furubayashi, the late Dr K Nii, and the late Dr R Nakagawa. Much of the authors' early x-ray imaging work was done at Japan's synchrotron facility, the Photon Factory, KEK, while the authors later decided to move to the laboratory source. Kenji Sakurai has been a user at the Photon Factory since 1983, the first year of operation. The authors thank the beamline scientists at BL-4A, BL-14B, BL-16A1 and AR-NW2A for their assistance during the experiments. The authors have enjoyed discussions on x-ray spectroscopy and imaging with the Japanese x-ray science community, especially the *Discussion Group of X-Ray Analysis, The Japan Society for Analytical Chemistry*. The authors are also grateful to many leading international scientists who have organized important scientific platforms, such as the *Denver X-ray Conference*, the *European X-ray Spectrometry Conference*, the *Total-Reflection X-ray Fluorescence Conference*, the *X-ray Optics and Microscopy Conference*, and Wiley's *X-ray Spectrometry Journal*. Many of their important and valuable works are cited as references in this book.

The authors would like to thank Dr John Navas and Dr Kosei Kameta of IOP Publishing for their excellent and patient editorial support.

Author biographies

Kenji Sakurai

Kenji Sakurai, born in 1959, received his doctorate in 1988 from The University of Tokyo (Professor Yohichi Gohshi's laboratory). During his master's and PhD research, he had the opportunity to gain invaluable knowledge and expertise under the guidance of Professor Atsuo Iida at Japan's synchrotron facility, Photon Factory, KEK. In 1988, he began his tenure at the National Institute for Metals (later reorganized as the National Institute for Materials Science, NIMS), Tsukuba, Japan. From 1993 to 1994, he was a visiting scientist in the laboratory of Professor Keith Bowen at the University of Warwick in the United Kingdom. Since 2004, in addition to his position at NIMS, he has also served as a professor at the University of Tsukuba. For 16 years he gave a weekly lecture, Introduction to X-ray Physics, in English mainly to international students. He continued a joint monthly colloquium with the laboratory of Professor Sadao Aoki at the university. Currently, he is an honorary researcher at NIMS and a professor emeritus at University of Tsukuba after his retirement in March 2020. His research interests include novel analytical imaging methods and instrumentation using x-rays and neutrons, (especially those used to solve problems related to surfaces and interfaces) as well as research into non-crystalline solids and some inorganic crystals. He has been a scientific programmer for more than 40 years (Delphi, Pascal, C++, Basic, Fortran, etc.). He has served as an editorial board member of the *Journal of Physics: Condensed Matter* (2008–2014) and News Editor of *X-ray Spectrometry* (2005–2024). He maintains a private laboratory, the Imaging Physics Laboratory, in Tokai, Ibaraki, near the nuclear reactor and large accelerator facility. Its address is: Imaging Physics Laboratory, Ibaraki Neutron Medical Research Center E301, 162-1, Shirakata, Tokai, Ibaraki, 319-1106 Japan. His recent activities are listed on the following web pages:

https://researchmap.jp/kenji.sakurai.xray?lang=en
https://xray-neutron-buried-interface.jp/lab/backgrounde.html

Wenyang Zhao

Wenyang Zhao, born in 1993, received his bachelor's degree in materials science and engineering from Tsinghua University in 2014. He then joined Professor Kenji Sakurai's X-ray laboratory through a joint program between the National Institute for Materials Science and the University of Tsukuba, specializing in advanced x-ray analysis and imaging techniques. In 2020, he received his doctoral degree in materials science. After that, he joined the Institute of Physical and Chemical Research (RIKEN), and he is currently a postdoctoral researcher at the RIKEN Center for Computational Science. His research focuses on utilizing high-performance computing to analyze diverse microscopy images, including x-ray free-electron laser imaging data.

Chapter 1

The colors of the chemical elements

1.1 Seeing is believing

'Seeing is believing' [1] is a common phrase implying that people are more likely to believe something as the truth if they see it with their own eyes. This idea has been particularly relevant through the history of science, where direct and reproducible observations and measurements have provided reliable evidence supporting ideas, hypotheses, and theories. In addition to supporting existing ideas, observations and measurements have often led to new discoveries resulting in changes, reconstructions, and even revolutions in established academic knowledge at the time.

In science, seeing is not always limited to what can be seen with the naked eye. Scientists use various instruments and tools such as microscopes, telescopes, and imaging technologies. These tools have been invented and developed to see things that would otherwise be too small, too far away, or too difficult to detect due to lack of sensitivity. In early pathology, biologists and medical doctors struggled because they lacked tools to see tiny microbes smaller than the spatial resolution of optical microscopes. Viruses are submicroscopic infectious agents that replicate only inside living cells. They are now well known because of sensitive electron microscopes that can visualize them. In 1939, the tobacco mosaic virus was first observed directly by electron microscope [2].

Even in the 21st century, developing new measurement methods, instruments, and technologies continues to play a crucial role in scientific breakthroughs. The James Webb Space Telescope, named the 2022 Breakthrough of the Year by Science magazine [3], demonstrates the importance of the 'seeing is believing' principle in science. For example, as shown in figure 1.1, the new telescope has provided much clearer details of the dwarf galaxy Wolf–Lundmark–Melotte. Using its ability to image in the infrared range, this telescope has captured valuable images that were previously impossible to obtain. One reason for such great success of the telescope was the full use of the latest knowledge and technology in materials engineering, including lightweight, high-strength infrared mirrors that can withstand temperatures as low as −240 °C [4]. By introducing cutting-edge knowledge and technology into measuring equipment, scientists have made significant advancements and explored unknown

doi:10.1088/978-0-7503-3215-6ch1

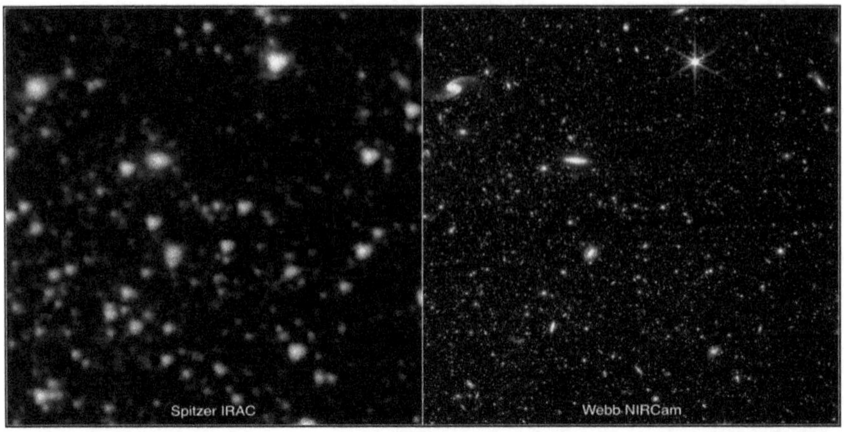

Figure 1.1. A portion of the dwarf galaxy Wolf–Lundmark–Melotte (WLM) captured by the Spitzer Space Telescope's Infrared Array Camera (left) and the James Webb Space Telescope's Near-Infrared Camera (right) [5]. The images demonstrate Webb's remarkable ability to resolve faint stars outside the Milky Way. The Spitzer image shows 3.6-micron light in cyan and 4.5-micron in orange. (IRAC1 and IRAC2). The Webb image includes 0.9-micron light shown in blue, 1.5-micron in cyan, 2.5-micron in yellow, and 4.3-micron in red (filters F090W, F150W, F250M, and F430M). Credits for science: NASA, ESA, CSA, IPAC, Kristen McQuinn (RU). Credits for image processing: Zolt G. Levay (STScI), Alyssa Pagan (STScI). Left: Image credit, Spitzer Space Telescope's Infrared Array Camera and NASA. Right: James Webb Space Telescope's Near-Infrared Camera and NASA.

areas. The James Webb Space Telescope's success is an excellent example that highlights the significance of developing advanced measurement methods, equipment, and technology to see what was previously invisible in scientific research.

Electron microscopes and infrared space telescopes are not the only tools that help us see and understand nature better. Because scientists need all kinds of additional sophisticated tools to see the invisible, it is not enough to rely on existing instruments. It is important to develop new instruments that take full advantage of the latest knowledge and technology. The authors of this book intend to provide complete information on x-ray color imaging, which will help scientists to solve many remaining scientific mysteries, by visualizing chemical elements. Scientists will see and believe nature through the colors of the chemical elements. The method is extremely powerful, and therefore it will be possible to understand new scientific discoveries with more confidence.

1.2 Chemical elements and the periodic table—a modern way to see nature

1.2.1 Mendeleev's periodic table

When exploring nature, understanding chemical composition is important. It is crucial to understand not only the shape and surface color but also the chemical elements that compose matter. The periodic table is a fundamental tool in science that helps organize and categorize elements based on their physical and chemical properties. Figure 1.2 shows the modern periodic table [6].

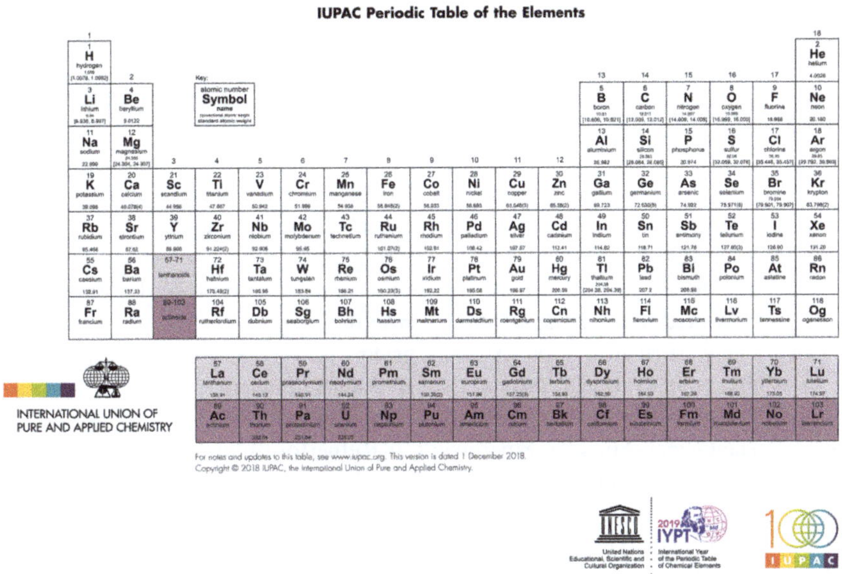

Figure 1.2. The modern periodic table [6]. The International Union of Pure and Applied Chemistry officially recognizes a total of 118 chemical elements (as at March 2023). The first 94 occur naturally on Earth, and the remaining 24 are synthetic elements produced in nuclear reactions.

The periodic table organizes chemical elements by atomic number and electron configuration. This systematic framework allows scientists to predict the physical and chemical properties of an element under various conditions. The periodic table also helps us to understand how elements behave in chemical reactions. As chemical elements in the same group have similar properties, the periodic table gives guidelines for the synthesis of new compounds and materials with specific properties and functions. This idea is widely used in designing new materials for electronics, energy, medicine, and other fields. In short, the periodic table is an essential tool in almost all scientific fields, such as chemistry, physics, biology, engineering, materials science, and more.

This useful table has a long history. In fact, the history of the periodic table is the history of chemistry [7]. However, the Russian chemist Dmitri Mendeleev (1834–1907) made a definitive breakthrough. In 1869, Mendeleev arranged 63 chemical elements by atomic weight, noting recurring chemical properties [8]. He made updates in 1870 and 1871 [9]. At the time, the form of the periodic table was unclear, but Mendeleev's version led to the prediction of the then-undiscovered chemical elements such as gallium, scandium, and germanium, found in 1875, 1879, and 1886, respectively.

Over the decades, our understanding of the chemical elements progressed by clarifying their atomic structure and using x-rays. Scientists learned that atoms have a positive nucleus and negative electrons. Table 1.1 summarizes the pioneering work in atomic physics that took place in the early 20th century. Although Mendeleev's table was arranged by atomic weight, new atomic knowledge showed the importance of the atomic number. This research was done by groups such as

Table 1.1. The progress of atomic physics in early 20th century.

1895	The discovery of x-rays (Wilhelm Röntgen)
1897	The discovery of negatively charged subatomic particles (currently recognized as electrons) in cathode rays (Joseph John Thomson)
1901	The Nobel Prize in Physics was awarded for x-rays (Wilhelm Röntgen)
1901	Pioneering work related to x-ray fluorescence spectra (Georges Sagnac)
1904	Thomson's plum pudding model (Joseph John Thomson)
1904	Nagaoka's Saturn model (Hantaro Nagaoka)
1905	The discovery of element-dependent secondary x-rays (Charles Barkla, who worked under Joseph John Thomson)
1905	*Annus Mirabilis* papers (Albert Einstein)
1906	The Nobel Prize for Physics (Joseph John Thomson)
1908	The Nobel Prize for Chemistry (Ernest Rutherford)
1908–13	The Geiger–Marsden experiments at Rutherford's lab
1911	Rutherford's nuclear model (Ernest Rutherford)
1912	The discovery of x-ray diffraction (Max von Laue)
1913	Bragg's law (Henry and Lawrence Bragg)
1913	Bohr's atomic model (Niels Bohr)
1913	Moseley's law (Henry Moseley)
1914	The Nobel Prize for Physics (Max von Laue)
1915	The Nobel Prize for Physics (Henry and Lawrence Bragg)
1917	The discovery of protons (Ernest Rutherford)
1917	The Nobel Prize for Physics (Charles Barkla)
1918	The Nobel Prize for Physics (Max Planck)
1920	The start of the hunt for neutrons (Ernest Rutherford, James Chadwick)
1921	The Nobel Prize for Physics (Albert Einstein)
1922	The Nobel Prize for Physics (Niels Bohr)
1923–25	de Broglie's equations for matter waves (Louis de Broglie)
1924	The Nobel Prize for Physics (Manne Siegbahn)
1926	The Schrödinger equation and quantum physics (Erwin Schrödinger)
1927	The uncertainty principle (Werner Heisenberg)
1927	The Bohr–Einstein debates at the Solvay conference
1927	The Nobel Prize for Physics (Arthur Compton)
1929	The Nobel Prize for Physics (Louis de Broglie)
1932	The Nobel Prize for Physics (Werner Heisenberg)
1932	Irène and Frédéric Joliot's experiment
1932	Discovery of neutrons (James Chadwick)
1933	The Nobel Prize for Physics (Erwin Schrödinger, Paul Dirac)
1935	The Nobel Prize for Physics (James Chadwick)

Joseph John Thomson's, Ernest Rutherford's, and other pioneering quantum physicists. As discussed later, their research was closely connected to x-ray spectroscopy, which explored chemical elements and filled gaps in the periodic table.

1.2.2 The discovery of x-rays

On November 8, 1895, x-rays were discovered at the University of Würzburg in Germany by Wilhelm Röntgen (1845–1923) [10]. He was doing experiments in a dark room using a Crookes tube, a vacuum tube with nearly 20 kV applied between the cathode and the anode (both made from platinum). To study cathode rays, the setup was shielded with black paper. This made the room dark, as the light of the tube's discharge was unseen. However, Röntgen noticed a faint light. At first, he thought the light came from the tube. But the tube was perfectly shielded, so that was impossible. He found the light came from a barium platinum cyanide ($Ba[Pt(CN)_4]$) fluorescent screen, accidentally left on the table. He concluded that invisible radiation (now called x-rays) emitted from the tube passed through the glass. Röntgen won the first Nobel Prize in Physics in 1901 for discovering x-rays.

At the time, x-rays' high transmission power was most interesting and impressive. Seventeen years after Röntgen's discovery, the next wave arrived. In 1912, Max von Laue (1879–1942) discovered x-ray diffraction by crystals [11]. Henry Bragg (1862–1960) and his son Lawrence Bragg (1890–1971) performed x-ray diffraction experiments in reflection geometry with their spectrometer, proposing Bragg's law [12]. Laue and the Braggs won Nobel Prizes in Physics in 1914 and 1915, respectively.

When x-rays hit matter, other x-rays are generated, called secondary x-rays. Most of these are now understood to be characteristic x-rays or x-ray fluorescence, whose wavelengths are intrinsic to the chemical elements contained in the matter. X-ray fluorescence was discovered before x-ray diffraction, then studied in more detail using x-ray spectrometers based on the x-ray diffraction phenomenon. After pioneering work [13–15] in 1901 by Georges Sagnac (1869–1928), x-ray fluorescence was discovered to be element-specific secondary x-rays by Charles Barkla (1877–1944) around 1903–1906 during work on x-ray polarization and angle-dependent scattering at Joseph John Thomson's lab [16–23]. By 1911, Barkla had found continuous spectra and a series of K- and L-series x-ray fluorescence lines in secondary x-rays [24–31]. Barkla assigned names to the x-ray fluorescence lines, such as $K\alpha$, $K\beta$, $L\alpha$, $L\beta$, etc. at that time. He received the Nobel Prize in Physics in 1917 for his discovery of the characteristic x-rays of the elements.

1.2.3 Moseley's law

A great breakthrough in x-ray fluorescence spectroscopy was due to Henry Moseley (1887–1915). As a young member of Ernest Rutherford's lab, Moseley systematically studied x-ray fluorescence spectra for almost all known chemical elements. After x-ray diffraction was discovered, Bragg's law enabled single crystals to be used as x-ray wavelength analyzers. Moseley contacted the Braggs and learned about their x-ray spectrometer. In his experiments, he used a fine crystal of potassium ferrocyanide ($K_3[Fe(CN)_6]\bullet 3H_2O$) with a good, large cleavage face as an analyzing crystal [32, 33]. By rotating the crystal, he found the angle satisfying the diffraction condition for x-ray fluorescence. He then determined the x-ray fluorescence wavelength from the Bragg angle. He collected the wavelengths of two K lines ($K\alpha$ and $K\beta$)

for light elements and three L lines (Lα, Lβ, and Lγ) for heavy elements. After sorting through all this data, he discovered a linear relationship between the atomic number and the square root of the x-ray frequency, known as Moseley's law [34, 35].

Figure 1.3 shows the 21st century version of Moseley's law. The abscissa uses x-ray energy instead of frequency because the use of x-ray energy is common in modern science. Both x-ray energy and frequency are proportional to the inverse of wavelength, giving the same linear plot. In Moseley's original work, he assumed a relationship such as $\nu = \nu_0 \, A \, (Z - b)^2$, where ν is the frequency of the x-ray fluorescence, ν_0 is the fundamental Rydberg frequency, and Z is the atomic number. Here, A and b are characteristic constants for each line: for example, A for the Kα line is $1/1^2 - 1/2^2 = 0.75$, A for the Lα line is $1/2^2 - 1/3^2 = 0.14$, etc. Moseley's finding clearly indicated that x-ray spectra could identify chemical elements.

Unfortunately, Moseley's work was terminated because he was killed in 1915 during World War I. On the other hand, Manne Siegbahn (1886–1978) developed a precise x-ray spectrometer and collected many high-resolution x-ray spectra, including soft x-rays and the vacuum ultraviolet region [37–41]. Siegbahn won the Nobel Prize in Physics in 1924 for discovery and research in x-ray spectroscopy.

1.2.4 The discovery of the missing chemical elements

Moseley's law and x-ray spectroscopy greatly helped to complete the periodic table. At that time, many scientists competed with each other to discover the missing chemical elements in the periodic table. In 1923, hafnium (Hf, $Z = 72$) was discovered by Niels Bohr's group in Copenhagen [42]. In the modern table, $Z = 72$ is in the fourth column and the sixth row. Vertically aligned elements have similar properties, so Bohr predicted $Z = 72$ would resemble zirconium ($Z = 40$), though some argued otherwise. Bohr's group assumed that zirconium ores contained $Z = 72$. George von Hevesy (Bohr's former student) and Dirk Coster measured the ore's x-ray fluorescence spectra, finding lines for $Z = 72$ in Moseley's diagram. They named it hafnium, Latin for Copenhagen.

In 1925, rhenium (Re, $Z = 75$) was found by Walter Noddack, Ida Tacke, and Otto Berg using x-ray spectroscopy to analyze platinum ore, columbite, gadolinite, and molybdenite [43]. It was later found to be the same element that a Japanese chemist, Masataka Ogawa, claimed to have discovered about 20 years earlier, in 1908. He named it nipponium, but mistakenly thought that his discovery was for the other missing element 43. Unfortunately, x-ray spectroscopy was not generally used at that time, and no one was able to reproduce his discovery [44, 45]. Element 75 was named rhenium after the river Rhine (Rhenius). Technetium (Tc, $Z = 43$) was discovered in 1937 at the University of Palermo by Carlo Perrier and Emilio Segrè through artificial synthesis using the cyclotron [46].

In 1939, Marguerite Perey found francium (Fr, $Z = 87$) [47]. In 1940, Dale Corson, Kenneth MacKenzie, and Segrè made astatine (At, $Z = 85$) by bombarding bismuth-209 with alpha particles [48]. Those two elements are radioactive and have rather short half-lives, leading to difficulties in identification.

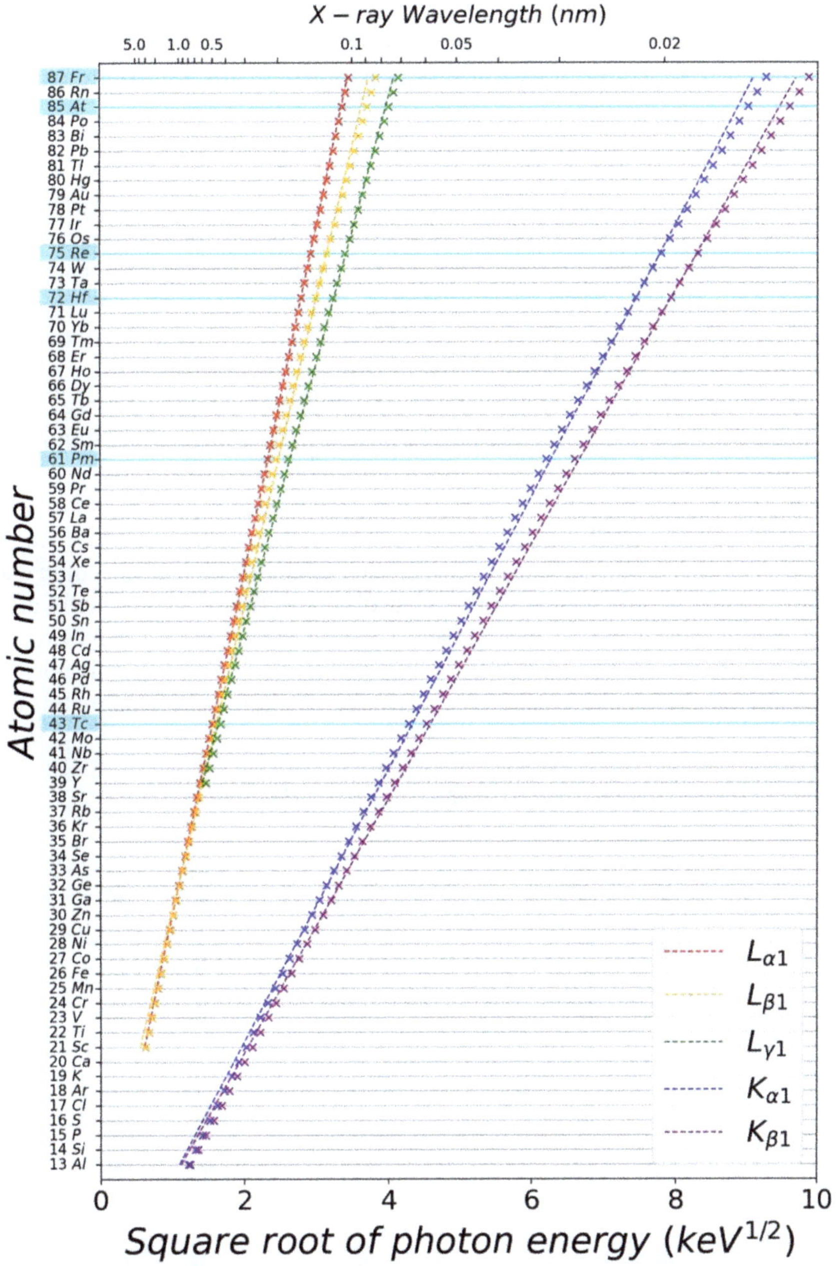

Figure 1.3. A modern drawing of Moseley's law, originally published in 1914 [35]. The abscissa has been changed from 'Square root of frequency' to 'Square root of photon energy.' In the replotting process, the modern numerical data library xraylib [36] was used. The relationship is almost linear, though some displacement is seen in the very high-Z part. The elements corresponding to $Z = 43$, 61, 72, 75, 85, and 87 were not known in 1914, but all six elements were discovered later with the help of Moseley's law.

In 1945, promethium (Pr, $Z = 61$) was produced and characterized at Oak Ridge National Laboratory by Jacob Marinsky, Lawrence Glendenin, and Charles Coryell, who studied uranium fission products produced in a graphite reactor [49].

1.3 The fundamentals of x-ray fluorescence

Following the pioneering physical research of the 20th century, x-ray fluorescence spectroscopy was reborn as a powerful, reliable chemical analysis method. Under sunlight or indoor lighting, many items in the environment reflect specific wavelengths corresponding to their colors, such as green, yellow, or red. Using these colors, it is possible to identify their optical properties. Similarly, under invisible x-ray irradiation, chemical elements contained in items emit x-ray fluorescence. Thanks to our knowledge of Moseley's law, the wavelengths (energy) of x-ray fluorescence allow us to identify chemical elements. Therefore, x-ray fluorescence is the color of a chemical element.

As shown in Figures 1.4 and 1.5, x-ray fluorescence occurs when primary x-rays are absorbed by atoms, exciting inner-shell electrons which are ejected. The energy of the primary x-rays must be higher than the absorption edge, which is the binding energy of the electrons. The ejected electrons are called photoelectrons. Immediately after such excitation, relaxation occurs to fill the vacancy in the orbital. In other words, energy equal to the difference in the energy levels of the transferred electrons is released into the environment. When this energy is emitted as an electromagnetic wave, it

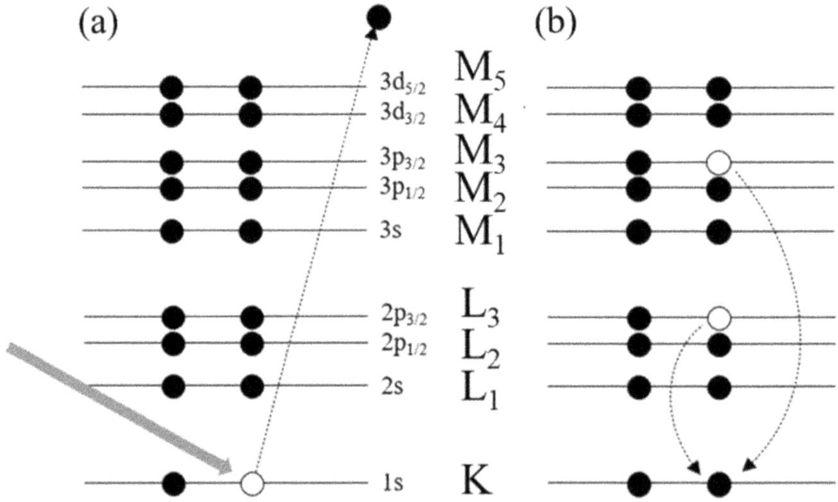

Figure 1.4. The physical principle of K-x-ray fluorescence. When the primary x-rays remove an electron from the K-shell (a), electrons in the L- or M-shells can transfer to fill the hole in the K-shell (b). X-ray fluorescence whose energy is equal to the difference between those two levels is then generated. If the transition is from the L_3 shell ($2p_{3/2}$ orbital) to the K-shell (1s orbital), $K\alpha_1$ x-ray fluorescence is generated. If the transition is from the M_3 shell ($3p_{3/2}$ orbital) to the K-shell (1s orbital), $K\beta_1$ x-ray fluorescence is generated.

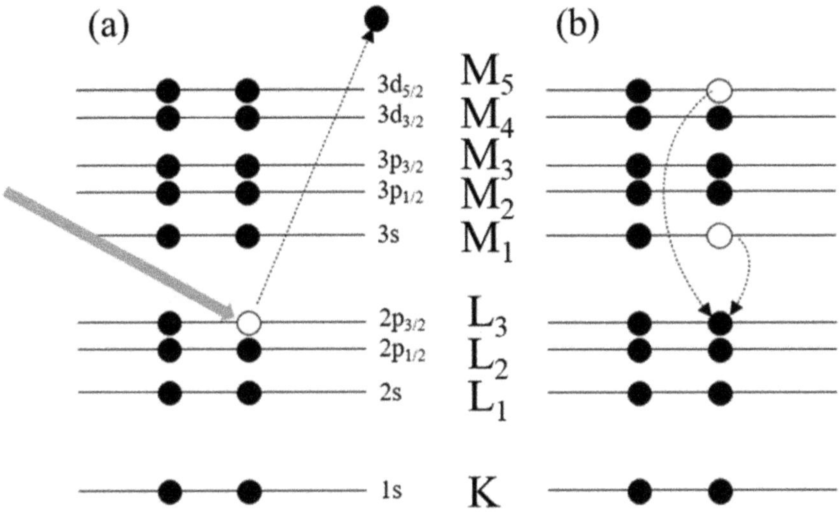

Figure 1.5. The physical principle of L-x-ray fluorescence. When the primary x-rays remove an electron from one of the L-shells (a), electrons in the M- (or N-) shells can transfer to fill the hole in the L-shell (b). X-ray fluorescence whose energy is equal to the difference between those two levels is then generated. If the transition takes place from the M_1 shell (3s orbital) to the L_3 shell ($2p_{3/2}$ orbital), L_l x-ray fluorescence is generated. If the transition takes place from the M_5 shell (3s orbital) to the L_3 shell ($2p_{3/2}$ orbital), $L\alpha_1$ x-ray fluorescence is generated.

corresponds to the generation of x-ray fluorescence. Since the relaxation processes are not single but multiple, the x-ray fluorescence energies are not single, resulting in the appearance of multiple lines in the x-ray spectra. The energies of such peaks correspond to the differences in the energy levels before and after the transitions.

Table 1.2 lists the names of the x-ray fluorescence lines. Due to the selection rule in dipole transitions, note that 2s → 1s, 3s → 1s, and 3d → 1s transitions are forbidden. As already mentioned, the wavelength data of x-ray fluorescence lines and x-ray absorption edges have been tabulated and provided in database [36]. Another useful database is available as a form of the periodic table [50]. See appendix A.

Since 1952 [51], theoretical x-ray fluorescence spectral models have been developed to predict x-ray fluorescence spectra for materials with any chemical composition. The Sherman [52–54] and Shiraiwa–Fujino [55, 56] formulas can quantitatively calculate and predict the x-ray fluorescence intensities produced by each element in a sample, leading to a breakthrough in this method of chemical element analysis [57–59]. When the samples include many different elements, it is necessary to consider the matrix effect. X-ray fluorescence intensity is sometimes enhanced by other x-ray fluorescence produced by other elements. X-ray fluorescence intensity is also sometimes weakened because of absorption by other elements in the same sample.

Table 1.3 is a list of the parameters used in the formula to calculate the x-ray fluorescence intensity. The formulas to be explained in this subsection are essentially the same as those in the literature reported by the pioneers [51–59]. Figure 1.6(a)

Table 1.2. The names of the x-ray fluorescence lines and the corresponding transitions between electron energy levels. Because the selection rule applies to dipole transitions, transitions such as 2s → 1s, 3s → 1s, and 3d → 1s are forbidden. This table only shows typical transitions. Many other x-ray fluorescence lines are produced by L_1, L_2, and L_3 excitation. Elements with large atomic numbers emit x-ray fluorescence due to M-shell excitation.

Shell	Name	Electron transition
K	$K\alpha_1$	$2p_{3/2}$ (L_3)→1s (K)
	$K\alpha_2$	$2p_{1/2}$ (L_2)→1s (K)
	$K\beta_1$	$3p_{3/2}$ (M_3)→1s (K)
	$K\beta_3$	$3p_{1/2}$ (M_2)→1s (K)
L_3	$L\alpha_1$	$3d_{5/2}$ (M_5)→$2p_{3/2}$ (L_3)
	$L\alpha_2$	$3d_{3/2}$ (M_4)→$2p_{3/2}$ (L_3)
	$L\beta_6$	4s (N_1)→$2p_{3/2}$ (L_3)
	$L\beta_{2,15}$	4d ($N_{4,\ 5}$)→$2p_{3/2}$ (L_3)
	L_l	3s (M_1)→$2p_{3/2}$ (L_3)
L_2	$L\beta_1$	$3d_{3/2}$ (M_4)→$2p_{1/2}$ (L_2)
	$L\gamma_1$	$4d_{3/2}$ (N_4)→$2p_{1/2}$ (L_2)
	$L\eta$	3s (M_1)→$2p_{1/2}$ (L_2)
L_1	$L\beta_3$	$3p_{3/2}$ (M_3)→2s (L_1)
	$L\beta_4$	$3p_{1/2}$ (M_2)→2s (L_1)
	$L\gamma_3$	$4p_{3/2}$ (N_3)→2s (L_1)
	$L\gamma_2$	$4p_{1/2}$ (N_2)→2s (L_1)

shows primary excitation, which is the simplest case. Suppose that monochromatic x-rays of wavelength λ have a cross-sectional area of 1 cm^2 and an intensity of I_0. Assuming that the primary x-ray is incident at an angle of Φ with respect to the sample surface, the intensity I_0 is attenuated before reaching a depth x_1. The x-ray fluorescence intensity p of element j per unit weight (here, x-rays such as $K\alpha_1$ and $L\beta_1$ are generally denoted by p) can now be expressed as the product of the absorption coefficient and other probabilities, and finally, the observed x-ray fluorescence intensity is obtained by integration along the depth to the thickness t [52, 55].

$$\begin{aligned} I_1(j,p) &= I_0 \frac{C_j \tau_{j,\lambda} \rho}{\sin \Phi} \frac{r_{j,p}-1}{r_{j,p}} g_{j,p} \omega_{j,p} \frac{\mathrm{d}\Omega}{4\pi} \frac{\sin \Phi}{\sin \Psi} \int_0^t \exp\left[-\rho x_1 \left(\frac{\mu_{S,\lambda}}{\sin \Phi} + \frac{\mu_{S,\lambda_j}}{\sin \Psi}\right)\right] \mathrm{d}x_1 \\ &= I_0 C_j \tau_{j,\lambda} \rho \frac{r_{j,p}-1}{r_{j,p}} g_{j,p}\, \omega_{j,p} \frac{\mathrm{d}\Omega}{4\pi} \frac{1}{\sin \Psi} \frac{1-\exp\left[-\rho t\left(\frac{\mu_{S,\lambda}}{\sin \Phi} + \frac{\mu_{S,\lambda_j}}{\sin \Psi}\right)\right]}{\frac{\mu_{S,\lambda}}{\sin \Phi} + \frac{\mu_{S,\lambda_j}}{\sin \Psi}} \end{aligned} \quad (1.1)$$

Table 1.3. A list of definitions of the parameters used in the theoretical expression for x-ray fluorescence intensity.

The intensity of the x-ray fluorescence *p* ($K\alpha_1$, $L\alpha_1$, etc.) of the target element *j* generated by the primary excitation	$I_1(j,p)$
The wavelength of the primary x-ray (assumed here to be monochromatic)	λ
The concentration of the target element *j*	C_j
The wavelength of the x-ray fluorescence *p* of the target element *j*	$\lambda_{j,p}$
The mass absorption coefficient of the sample with respect to the primary x-rays (sometimes called the mass attenuation coefficient)	$\mu_{S,\lambda}$
The mass absorption coefficient (sometimes called the mass attenuation coefficient) of the sample for the x-ray fluorescence *p* (wavelength $\lambda_{j,p}$) produced by the target element *j*	$\mu_{S,\lambda_{j,p}}$
The mass photoelectric absorption coefficient (the mass absorption coefficient in the strict sense) of the target element *j* for primary x-rays (of wavelength λ)	$\tau_{j,\lambda}$
The jump ratio at the absorption edge of the shell (K, L_3, L_2, L_1, etc.) excited during the generation of x-ray fluorescence *p* by the target element *j*	$r_{j,p}$
The transition probability of the target element *j* during the generation of x-ray fluorescence *p* (i.e. the probability of an electron transition from $2p_{3/2}$, $2p_{1/2}$, $3p_{3/2}$, etc. to 1s when a vacancy is created at 1s, taking K-shell excitation as an example)	$g_{j,p}$
The fluorescence yield of target element *j* (for the shell belonging to x-ray fluorescence *p*)	$\omega_{j,p}$
The density of the sample	ρ
The angle between the primary x-ray and the sample surface	Φ
The angle formed by the sample surface when the fluorescent x-ray of the target element *j* escapes from the sample	Ψ
The solid angle of the detector	$\Omega/4\pi$
The intensity of fluorescent x-ray *p* of the target element *j* generated by secondary excitation (by fluorescent x-ray *pp* of another element *jj*)	$I_2(j,p)$
The concentration of element *jj*	C_{jj}
The wavelength of the x-rays contributing to the first stage of secondary excitation (fluorescent x-rays *pp* of another element *jj* generated by primary x-rays)	$\lambda_{ii,pp}$
The mass absorption attenuation coefficient (sometimes called the mass attenuation coefficient) of the sample relative to the x-rays contributing to the secondary excitation (wavelength $\lambda_{jj,pp}$)	$\mu_{S,\lambda_{jj,pp}}$
The mass photoelectric absorption coefficient (the mass absorption coefficient in the strict sense) of element *jj* for the primary x-rays (wavelength λ)	$\tau_{jj,\lambda}$
The jump ratio at the absorption edge of the shell (K, L_3, L_2, L_1, etc.) excited during the generation of x-ray fluorescence *pp* by target element *jj*	$r_{jj,pp}$
The transition probability of element *jjj* during the generation of x-ray fluorescence *pp* (a breakdown of whether the transition occurs from 2p or 3p when a vacancy is created in 1s, taking K-shell excitation as an example)	$g_{jj,pp}$

Table 1.3. (*Continued*)

The fluorescence yield of element *jj* (for shells associated with the x-ray fluorescence *pp*)	$\omega_{jj,pp}$
The intensity of the fluorescence x-ray *p* of target element *j* generated by third-order excitation (by fluorescence x-rays *ppp*, *pp* of another element *jjj*, *jj*)	$I_3(j,p)$
The concentration of element *jjj*	C_{jjj}
The wavelength of x-rays contributing to the first stage of tertiary excitation (fluorescent x-rays *ppp* of another element *jjj*, generated by the primary x-rays)	$\lambda_{ijj,ppp}$
The mass absorption coefficient (sometimes called the mass attenuation coefficient) of the sample with respect to the x-rays (wavelength $\lambda_{jjjj,ppp}$) that contribute to the first stage of tertiary excitation	$\mu_{S,\lambda\ jjj,ppp}$
The mass photoelectric absorption coefficient (the mass absorption coefficient in the strict sense) of element *jjj* for the primary x-rays (wavelength λ)	$\tau_{jjj,\lambda}$
The jump ratio at the absorption edge of the shell (K, L_3, L_2, L_1, etc.) excited during the generation of x-ray fluorescence *ppp* by element *jjj*	$r_{jjj,ppp}$
The transition probability of element *jjj* during the generation of x-ray fluorescence *ppp* (a breakdown of whether the transition occurs from 2p or 3p when a vacancy is created in 1s, taking K-shell excitation as an example)	$g_{jjj,ppp}$
The fluorescence yield of element *jjj* (for shells belonging to the x-ray fluorescence *ppp*)	$\omega_{jjj,ppp}$

When the sample is very thick, $t \to \infty$.

$$I_1(j,p) = I_0 C_j \tau_{j,\lambda} \rho \frac{r_{j,p}-1}{r_{j,p}} g_{j,p} \omega_{j,p} \frac{\mathrm{d}\Omega}{4\pi} \frac{1}{\sin\Psi} \frac{1}{\dfrac{\mu_{S,\lambda}}{\sin\Phi} + \dfrac{\mu_{S,\lambda_j}}{\sin\Psi}} \tag{1.2}$$

When the sample is thin, $1-e^{-t} \to t$.

$$I_1(j,p) = I_0 C_j \tau_{j,\lambda} \rho \frac{r_{j,p}-1}{r_{j,p}} g_{j,p} \omega_{j,p} \frac{\mathrm{d}\Omega}{4\pi} \frac{1}{\sin\Psi} \rho t \tag{1.3}$$

Just for simplicity, excitation by monochromatic x-rays of wavelength λ is assumed. On the other hand, the primary x-rays produced by the usual commercially available laboratory x-ray sources are not purely monochromatic. Such x-ray sources provide continuous x-ray spectra in addition to characteristic x-ray lines (such as $K\alpha_1$, $K\alpha_2$, $K\beta_1$ … and *L* lines). To modify the spectral shape to obtain a better signal-to-background ratio, filtering is often employed. In such cases, all the above formulas require integration over the wavelength, considering the spectral shape of the primary x-rays. Alternatively, the 'effective wavelength' is often used in the above formula as a practical method.

Figure 1.6(b) and (c) show the cases for secondary and ternary excitation, respectively. They may appear complicated, but the calculation scheme itself is very similar to the primary excitation case. The secondary and ternary excitation are due to the x-ray fluorescence from coexisting elements in the sample.

The contribution of the secondary excitation (illustrated in figure 1.6(b)) can be expressed as follows [55].

$$
\begin{aligned}
I_2(j,p) &= I_0\left(C_{jj}\tau_{jj,\lambda}\rho\frac{r_{jj,pp}-1}{r_{jj,pp}}g_{jj,pp}\omega_{jj,pp}\right)\left(C_j\tau_{j,\lambda_{jj}}\rho\frac{r_{j,p}-1}{r_{j,p}}g_{j,p}\omega_{j,p}\right)\frac{\mathrm{d}\Omega}{4\pi}\frac{1}{2\sin\Psi}\times F\\
F &= \int_0^{\frac{\pi}{2}}\int_{x_2=0}^{t}\int_{x_1=0}^{x_2} Exp\left(-\left(\frac{\mu_{S,\lambda_j}}{\sin\Psi}+\frac{\mu_{S,\lambda_{jj}}}{\cos\theta}\right)\rho x_2\right)Exp\left(-\left(\frac{\mu_{S,\lambda}}{\sin\Phi}-\frac{\mu_{S,\lambda_{jj}}}{\cos\theta}\right)\rho x_1\right)\tan\theta dx_1 dx_2\, d\theta\\
&+\int_{\frac{\pi}{2}}^{\pi}\int_{x_2=0}^{t}\int_{x_1=x_2}^{t} Exp\left(-\left(\frac{\mu_{S,\lambda_j}}{\sin\Psi}-\frac{\mu_{S,\lambda_{jj}}}{\cos\theta}\right)\rho x_2\right)Exp\left(-\left(\frac{\mu_{S,\lambda}}{\sin\Phi}+\frac{\mu_{S,\lambda_{jj}}}{\cos\theta}\right)\rho x_1\right)\tan\theta dx_1 dx_2\, d\theta\\
&= \int_0^{\frac{\pi}{2}}\int_{x_2=0}^{t} Exp\left(-\left(\frac{\mu_{S,\lambda_j}}{\sin\Psi}+\frac{\mu_{S,\lambda_{jj}}}{\cos\theta}\right)\rho x_2\right)\frac{1-Exp\left(-\left(\frac{\mu_{S,\lambda}}{\sin\Phi}-\frac{\mu_{S,\lambda_{jj}}}{\cos\theta}\right)\rho x_2\right)}{\left(\frac{\mu_{S,\lambda}}{\sin\Phi}-\frac{\mu_{S,\lambda_{jj}}}{\cos\theta}\right)\rho}\tan\theta dx_2\, d\theta\\
&+\int_{\frac{\pi}{2}}^{\pi}\int_{x_2=0}^{t} Exp\left(-\left(\frac{\mu_{S,\lambda_j}}{\sin\Psi}-\frac{\mu_{S,\lambda_{jj}}}{\cos\theta}\right)\rho x_2\right)\frac{1-Exp\left(-\left(\frac{\mu_{S,\lambda}}{\sin\Phi}+\frac{\mu_{S,\lambda_{jj}}}{\cos\theta}\right)\rho x_2\right)}{\left(\frac{\mu_{S,\lambda}}{\sin\Phi}+\frac{\mu_{S,\lambda_{jj}}}{\cos\theta}\right)\rho}\tan\theta dx_2 d\theta\\
&= \int_0^{\frac{\pi}{2}}\left(\frac{1-Exp\left(-\left(\frac{\mu_{S,\lambda_j}}{\sin\Psi}+\frac{\mu_{S,\lambda_{jj}}}{\cos\theta}\right)\rho t\right)}{\left(\frac{\mu_{S,\lambda_j}}{\sin\Psi}+\frac{\mu_{S,\lambda_{jj}}}{\cos\theta}\right)\left(\frac{\mu_{S,\lambda}}{\sin\Phi}-\frac{\mu_{S,\lambda_{jj}}}{\cos\theta}\right)}-\frac{1-Exp\left(-\left(\frac{\mu_{S,\lambda}}{\sin\Phi}+\frac{\mu_{S,\lambda_j}}{\sin\Psi}\right)\rho t\right)}{\left(\frac{\mu_{S,\lambda}}{\sin\Phi}+\frac{\mu_{S,\lambda_j}}{\sin\Psi}\right)\left(\frac{\mu_{S,\lambda}}{\sin\Phi}-\frac{\mu_{S,\lambda_{jj}}}{\cos\theta}\right)}\right)\tan\theta\, d\theta\\
&+\int_{\frac{\pi}{2}}^{\pi}\left(\frac{Exp\left(-\left(\frac{\mu_{S,\lambda}}{\sin\Phi}-\frac{\mu_{S,\lambda_{jj}}}{\cos\theta}\right)\rho t\right)-Exp\left(-\left(\frac{\mu_{S,\lambda}}{\sin\Phi}+\frac{\mu_{S,\lambda_j}}{\sin\Psi}\right)\rho t\right)}{\left(\frac{\mu_{S,\lambda_j}}{\sin\Psi}+\frac{\mu_{S,\lambda_{jj}}}{\cos\theta}\right)\left(\frac{\mu_{S,\lambda}}{\sin\Phi}-\frac{\mu_{S,\lambda_{jj}}}{\cos\theta}\right)}-\frac{1-Exp\left(-\left(\frac{\mu_{S,\lambda}}{\sin\Phi}+\frac{\mu_{S,\lambda_j}}{\sin\Psi}\right)\rho t\right)}{\left(\frac{\mu_{S,\lambda}}{\sin\Phi}+\frac{\mu_{S,\lambda_j}}{\sin\Psi}\right)\left(\frac{\mu_{S,\lambda}}{\sin\Phi}-\frac{\mu_{S,\lambda_{jj}}}{\cos\theta}\right)}\right)\tan\theta d\theta
\end{aligned}
\tag{1.4}
$$

When the sample is very thick, $t \to \infty$,

$$
\begin{aligned}
I_2(j,p) &= I_0\left(C_{jj}\tau_{jj,\lambda}\rho\frac{r_{jj,pp}-1}{r_{jj,pp}}g_{jj,pp}\omega_{jj,pp}\right)\left(C_j\tau_{j,\lambda_{jj}}\rho\frac{r_{j,p}-1}{r_{j,p}}g_{j,p}\omega_{j,p}\right)\frac{\mathrm{d}\Omega}{4\pi}\frac{1}{2\sin\Psi}\times F\\
F &= \frac{1}{\frac{\mu_{S,\lambda}}{\sin\Phi}+\frac{\mu_{S,\lambda_j}}{\sin\Psi}}\left[\frac{\sin\Psi}{\mu_{S,\lambda_j}}\ln\left(1+\frac{\frac{\mu_{S,\lambda_j}}{\sin\Psi}}{\mu_{S,\lambda_{jj}}}\right)+\frac{\sin\Phi}{\mu_{S,\lambda}}\ln\left(1+\frac{\frac{\mu_{S,\lambda}}{\sin\Phi}}{\mu_{S,\lambda_{jj}}}\right)\right]
\end{aligned}
\tag{1.5}
$$

In the same way, the contribution of the ternary excitation (illustrated in figure 1.6(c)) can be expressed as follows [55]:

$$
\begin{aligned}
I = I_0 & \left(C_{jjj} \tau_{jjj,\lambda} \rho \frac{r_{jjj,ppp} - 1}{r_{jjj,ppp}} g_{jjj,ppp} \omega_{jjj,ppp} \right) \left(C_{jj} \tau_{jj,\lambda_{jjj}} \rho \frac{r_{jj,pp} - 1}{r_{jj,pp}} g_{jj,pp} \omega_{jj,pp} \right) \left(C_{j} \tau_{j,\lambda_{jj}} \rho \frac{r_{j,p} - 1}{r_{j,p}} g_{j,p} \omega_{j,p} \right) \\
& \times \frac{\mathrm{d}\Omega}{4\pi} \frac{1}{4 \sin \Psi} \frac{1}{\frac{\mu_{\mathrm{S},\lambda}}{\sin \Phi} + \frac{\mu_{\mathrm{S},\lambda_j}}{\sin \Psi}} \times F \\
F = & \left(\frac{\sin \Phi}{\mu_{\mathrm{S},\lambda}} \right)^2 \ln \left(1 + \frac{\frac{\mu_{\mathrm{S},\lambda}}{\sin \Phi}}{\mu_{\mathrm{S},\lambda_{jjj}}} \right) \ln \left(1 + \frac{\frac{\mu_{\mathrm{S},\lambda}}{\sin \Phi}}{\mu_{\mathrm{S},\lambda_{jj}}} \right) \\
& + \left(\frac{\sin \Phi}{\mu_{\mathrm{S},\lambda}} \right) \left(\frac{\sin \Psi}{\mu_{\mathrm{S},\lambda_j}} \right) \ln \left(1 + \frac{\frac{\mu_{\mathrm{S},\lambda}}{\sin \Phi}}{\mu_{\mathrm{S},\lambda_{jjj}}} \right) \ln \left(1 + \frac{\frac{\mu_{\mathrm{S},\lambda_j}}{\sin \Psi}}{\mu_{\mathrm{S},\lambda_{jj}}} \right) \\
& + \left(\frac{\sin \Psi}{\mu_{\mathrm{S},\lambda_j}} \right)^2 \ln \left(1 + \frac{\frac{\mu_{\mathrm{S},\lambda_j}}{\sin \Psi}}{\mu_{\mathrm{S},\lambda_{jjj}}} \right) \ln \left(1 + \frac{\frac{\mu_{\mathrm{S},\lambda_j}}{\sin \Psi}}{\mu_{\mathrm{S},\lambda_{jj}}} \right) \\
& + \left(\frac{\sin \Phi}{\mu_{\mathrm{S},\lambda}} + \frac{\sin \Psi}{\mu_{\mathrm{S},\lambda_j}} \right) \left(\frac{1}{\mu_{\mathrm{S},\lambda_{jj}}} \ln \left(1 + \frac{\mu_{\mathrm{S},\lambda_{jj}}}{\mu_{\mathrm{S},\lambda_{jjj}}} \right) + \frac{1}{\mu_{\mathrm{S},\lambda_{jjj}}} \ln \left(1 + \frac{\mu_{\mathrm{S},\lambda_{jjj}}}{\mu_{\mathrm{S},\lambda_{jj}}} \right) \right) \\
& - \frac{\sin \Phi}{\mu_{\mathrm{S},\lambda}} \int_0^{\frac{\mu_{\mathrm{S},\lambda_{jjj}}}{\mu_{\mathrm{S},\lambda_{jj}}}} \left[\frac{1}{\frac{\mu_{\mathrm{S},\lambda} t}{\sin \Phi} + \mu_{\mathrm{S},\lambda_{jjj}}} \ln \left(\frac{1+t}{t} \right) \right] \mathrm{d}t \\
& - \frac{\sin \Psi}{\mu_{\mathrm{S},\lambda_j}} \int_0^{\frac{\mu_{\mathrm{S},\lambda_{jj}}}{\mu_{\mathrm{S},\lambda_{jjj}}}} \left[\frac{1}{\frac{\mu_{\mathrm{S},\lambda_j} t}{\sin \Psi} + \mu_{\mathrm{S},\lambda_{jj}}} \ln \left(\frac{1+t}{t} \right) \right] \mathrm{d}t
\end{aligned}
\tag{1.6}
$$

All the calculations discussed above assume that the sample is uniform in the depth direction. On the other hand, it is sometimes necessary to handle a layered system. Figure 1.7 illustrates the secondary excitation in a two-layer system. As shown in figure 1.7(a), if the lower layer does not contribute to the matrix effect, the situation is essentially the same as in equation (1.4). On the other hand, if the lower layer contributes (as illustrated in figure 1.7(b)), it is necessary to consider such effects. Assuming that the thickness and the density of the upper layer are t_1 and ρ_1, respectively, and the thickness and the density of the lower layer are t_2 and ρ_2, respectively, the contribution can be calculated as follows [58].

$$
\begin{aligned}
I &= I_0 \left(C_{jj} \tau_{jj,\lambda} \rho \frac{r_{jj,pp} - 1}{r_{jj,pp}} g_{jj,pp} \omega_{jj,pp} \right) \left(C_{j} \tau_{j,\lambda_{jj}} \rho \frac{r_{j,p} - 1}{r_{j,p}} g_{j,p} \omega_{j,p} \right) \frac{\mathrm{d}\Omega}{4\pi} \frac{Exp\left(- \frac{\mu_{\mathrm{S1},\lambda} \rho_1 t_1}{\sin \Phi} \right)}{2 \sin \Psi} \times F \\
F &= \int_0^{\frac{\pi}{2}} \frac{Exp\left(- \frac{\mu_{\mathrm{S1},\lambda} \rho_1 t_1}{\cos \theta} \right) \left(1 - Exp\left(- \left(\frac{\mu_{\mathrm{S2},\lambda}}{\sin \Phi} + \frac{\mu_{\mathrm{S2},\lambda_j}}{\cos \theta} \right) \rho_2 t_2 \right) \right) \left(1 - Exp\left(- \left(\frac{\mu_{\mathrm{S1},\lambda_{jj}}}{\sin \Psi} - \frac{\mu_{\mathrm{S1},\lambda_j}}{\cos \theta} \right) \rho_1 t_1 \right) \right)}{\left(\frac{\mu_{\mathrm{S2},\lambda}}{\sin \Phi} + \frac{\mu_{\mathrm{S2},\lambda_j}}{\cos \theta} \right) \left(\frac{\mu_{\mathrm{S1},\lambda_{jj}}}{\sin \Psi} - \frac{\mu_{\mathrm{S1},\lambda_j}}{\cos \theta} \right)} \tan \theta \, d\theta
\end{aligned}
\tag{1.7}
$$

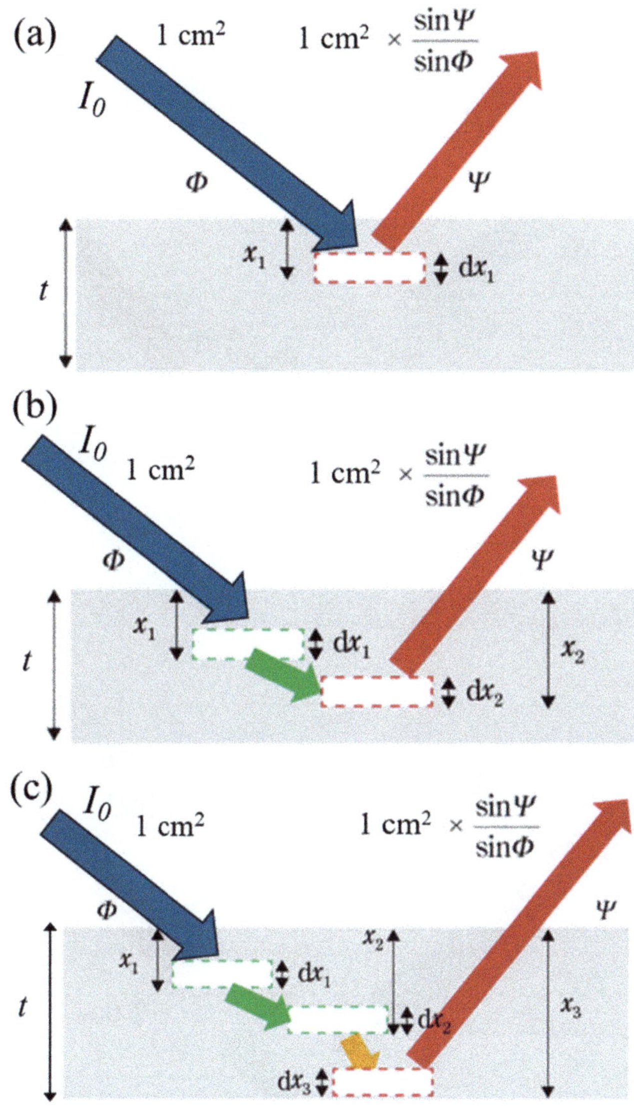

Figure 1.6. (a) A schematic illustration of x-ray fluorescence produced by primary excitation. (b) A schematic illustration of x-ray fluorescence produced by secondary excitation. (c) A schematic illustration of x-ray fluorescence produced by ternary excitation.

Although the above formula is for a simple bilayer system, the calculation can easily be extended to multilayer cases. The x-ray fluorescence intensity of a given element in each layer can be calculated in almost the same way by simply assuming the number of layers, the thickness of each layer, and the matrix composition. However, even if the calculation is possible, whether the actual structure of the measured sample is really like that or not is another matter. For practical applications, it is necessary to examine whether the sample has a layered structure with a sharp interface.

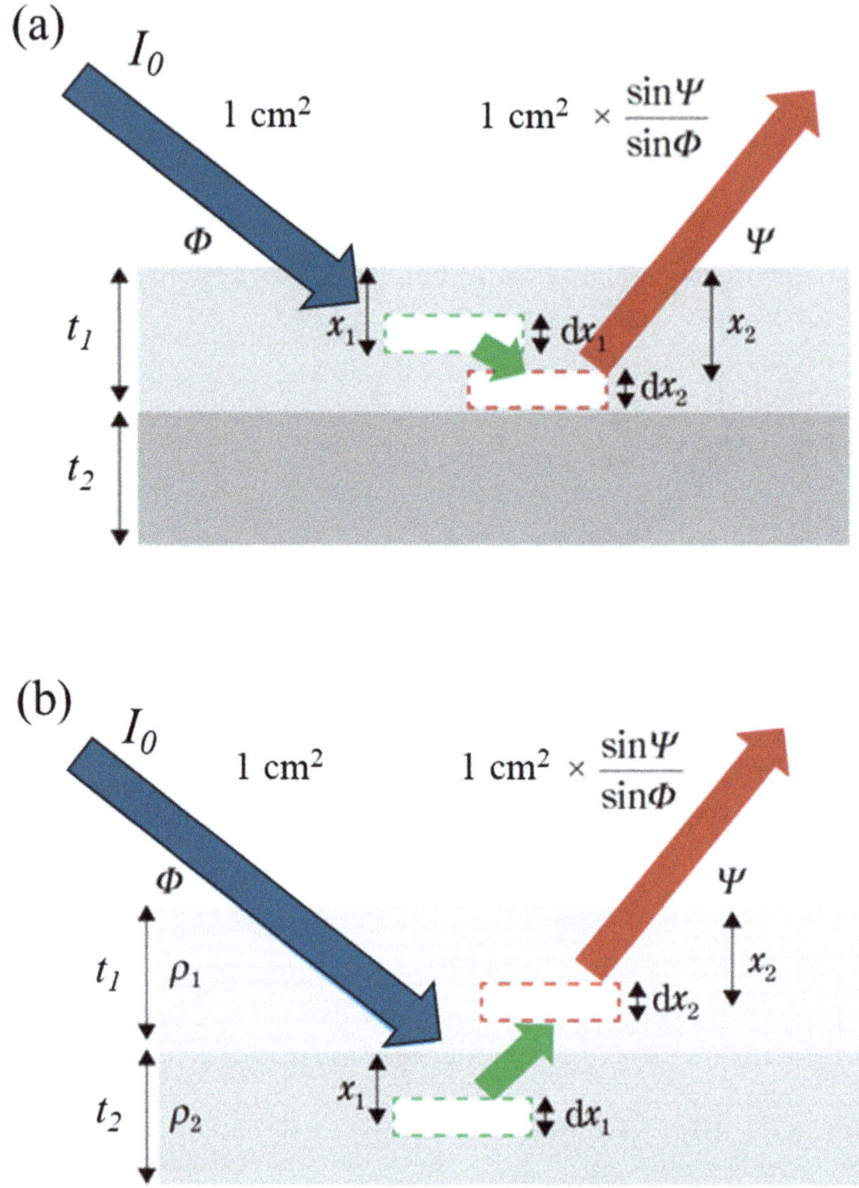

Figure 1.7. A schematic illustration of x-ray fluorescence produced by secondary excitation in the case of a two-layer system, without (a) and with (b) a contribution from the lower layer.

1.4 Energy-dispersive x-ray detectors

The advent of the energy-dispersive x-ray detector, which can resolve the x-ray energy of photons, was an extremely important development in seeing the colors of chemical elements [60–65], although the use of analyzing crystals based on x-ray diffraction had already been established in the early days. Such a detector uses the physical principle that semiconductors have bandgaps and that electron–hole pairs created by external radiation can be collected by applying a bias voltage. These detectors are also called semiconductor diode detectors or simply semiconductor detectors, although the world's first such detector was called a crystal counter. It used silver chloride (AgCl) and was made by P. J. van Heerden in 1943 and described in his PhD thesis in 1945 [66].

In 1949, Kenneth G. McKay developed the pioneering use of germanium in a radiation detector [67, 68]. Since then, germanium and silicon have been frequently employed for such detectors. Unlike conventional x-ray spectrometers based on analyzing crystals, radiation detector systems only use a single detector and an electronic circuit to process the pulse height, which corresponds to the difference in x-ray energy. Figure 1.8 schematically illustrates the principle of the energy-dispersive x-ray detector based on a p–n semiconductor diode with a depletion zone formed by a reverse bias voltage [63, 69].

Table 1.4 summarizes the properties of selected semiconductors which are promising candidates for use as radiation detectors [60, 63, 65]. Choosing materials with smaller ε, which is the energy required to create one electron–hole pair in the semiconductor device, is crucial, as the energy resolution is related to the statistical error in counting the number of electron–hole pairs created by a single x-ray photon. The value of ε is approximately given by $\varepsilon = 2.8\ E_g + \alpha$, where E_g is a bandgap, and

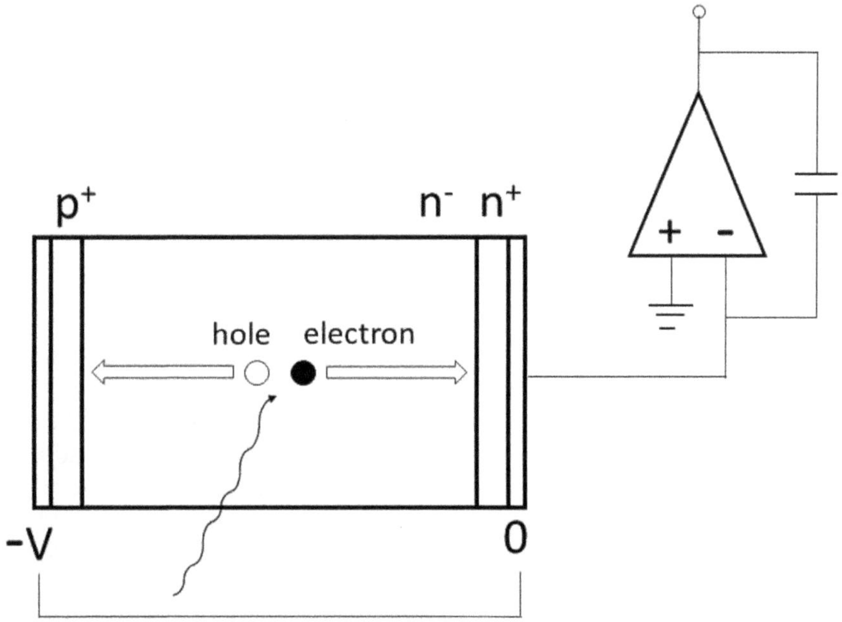

Figure 1.8. A simplified physical model of a p–n diode detector. The use of a reverse bias voltage depletes the semiconductor bulk and forms a drift field, leading to the collection of charges created by external radiation. Further details are described elsewhere [60, 69].

Table 1.4. The properties of semiconductors that are promising for radiation detectors. The detection efficiency, energy resolution, and throughput at high count rates are related to the number of electrons per volume (atomic number, *Z*, and density), the number of electron–hole pairs generated (bandgap and ε), and the charge collection time, respectively. The values are taken from sources [60, 63, 65].

					Mobility [cm^2 V^{-1} S^{-1}]	
	Atomic number(s) (*Z*)	Density [g/cm^3]	Bandgap [eV]	ε[eV]	Electron	Hole
Si	14	2.3	1.1	3.6	1350	480
Ge	32	5.3	0.7	3	3800	1800
Diamond	6	5.3	3.5	5.5	4500	3500
GaAs	31,35	5.4	1.5	4.2	8600	400
HgI_2	89,53	6.3	2.1	6.5	100	4
CdTe	48,52	6.1	1.5	4.43	1100	100
CdZnTe	48,30,52	5.3	1.5–2.4	5	1350	120

α is around 0.5~1 eV for germanium and silicon [70]. The theoretical energy resolution at the full width at half maximum (FWHM), ΔE at the energy *E*, can be expressed using ε as $\Delta E = 2.35\sqrt{F\varepsilon\ E}$, where *F* is a Fano factor [71]. *F* is around 0.11 for germanium and silicon. Other important criteria in the choice of a material are the electron density and the mobility, which are related to the detection efficiency and the charge collection time at high count rates, respectively. For higher-energy x-rays, such as 50–100 keV or even more, germanium is more suitable than silicon. Other materials such as CdTe and CdZnTe are also employed [60]. Diamond detectors are attractive for special fast measurements. High-quality single crystals of diamond are prepared using the chemical vapor deposition (CVD) technique [61]. HgI_2 is promising for operation at room temperature [65], while many other semiconductors usually require cooling.

To date, many different types of germanium and silicon detectors have been developed. As the lithium drift method [69] is an effective way to increase the depletion layer, leading to a large effective volume, both Ge(Li) [72] and Si(Li) detectors [73] have been developed and have been widely used. For germanium, Ge (Li) was later replaced by a high-purity device, because almost the same large effective volume became available.

In 1983, at Brookhaven National Laboratory, USA, Emilio Gatti, and Pavel Rehak invented a new concept for the semiconductor detector, the semiconductor drift detector [74, 75]. Figure 1.9 shows the structure of the modern silicon drift detector (SDD) [76–78]. In the 21st century, the SDD is almost the standard for energy-dispersive x-ray spectroscopy. Its great advantage is that it can manage both higher energy resolution and large effective area size, while conventional detectors have a limited capacity for areal enlargement because of the resulting reduction in the signal charge amount due to the increased area. The SDD can be operated with simple Peltier cooling and does not require liquid nitrogen like a Si(Li) detector. This is another practical advantage in the daily use of the energy-dispersive x-ray spectrometer.

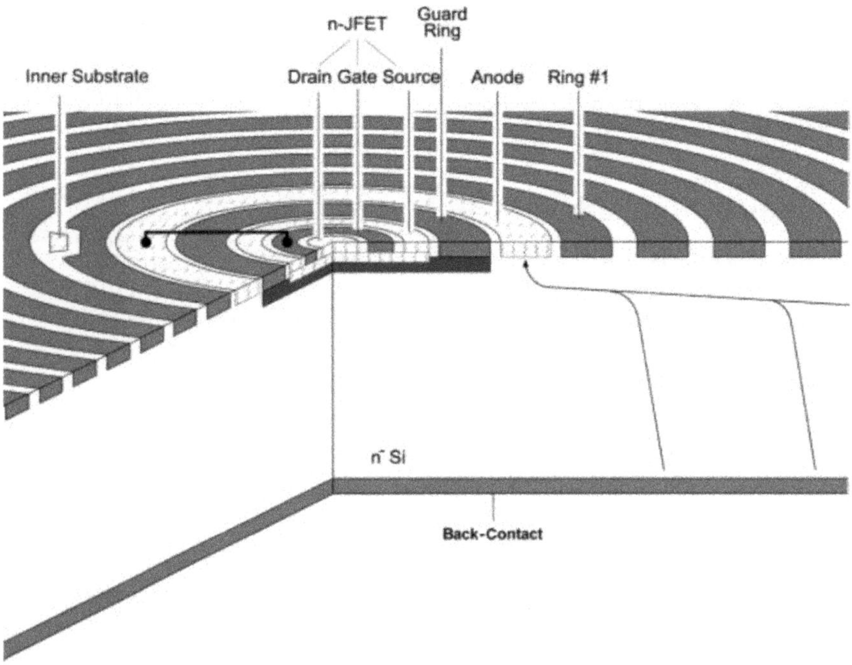

Figure 1.9. The typical structure of a silicon drift detector (SDD) [78], copyright 2004 John Wiley & Sons Ltd. The silicon is completely depleted and therefore the entire area is sensitive to incoming x-ray photons. An integrated field-effect transistor is placed in the center of the SDD. Electrons are guided by an electric field toward the anode ring, which is connected to the transistor gate by a metal strip.

1.5 Practical notes for modern x-ray fluorescence analysis

Nowadays, x-ray fluorescence spectroscopy is established as one of the most powerful and reliable chemical analysis tools in many scientific and engineering fields [79–83]. As already described, the advent of solid-state radiation detectors enabled another great invention, namely simultaneous multielement analysis. Such an x-ray spectrometer is called an energy-dispersive system. It differs from the conventional wavelength-dispersive system due to its use of analyzing crystals.

Figure 1.10 summarizes two different types of x-ray fluorescence spectrometers. The wavelength-dispersive type offers higher energy resolution and separates the $K\alpha_1$ and $K\alpha_2$ peaks, while the energy-dispersive type, based on semiconductor detectors (since mid-20th century), does not separate them except at very high energies for heavy elements. Overlapping $K\alpha_1$ and $K\alpha_2$ emissions are often denoted by just $K\alpha$. Overlapping $K\beta_1$ and $K\beta_3$ emissions are denoted by just $K\beta$. The energies of the $K\alpha$ and $K\beta$ peaks are between those of the $K\alpha_1$ and $K\alpha_2$ peaks and between those of the $K\beta_1$ and $K\beta_3$ peaks, respectively. Because of the low energy resolution of energy-dispersive spectrometers, these deviations are small.

Let us see what the x-ray fluorescence spectra of practical samples look like. Figure 1.11 shows an example of dried residue of horse serum [84]. Here, two

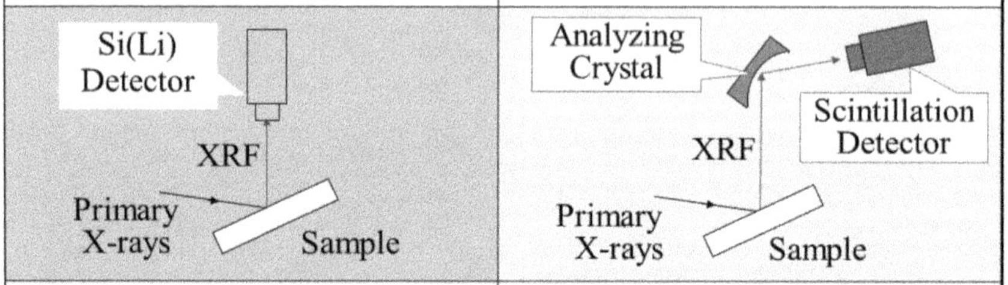

Energy Dispersive	Wavelength Dispersive
- **Simultaneous multi-element analysis** - High detection efficiency (large solid angle) - Low energy-resolution - Limitation of counting rate - Scattering background	- Usually requires some scans - Low detection-efficiency (requires high-photon flux) - High energy-resolution (availability of advanced analysis) - Good signal to background ratio

Figure 1.10. A comparison of two types of x-ray fluorescence spectrometers. The energy-dispersive type has been widely used since the mid-20th century. Its advantage is its capability for simultaneous multielement analysis, although the $K\alpha_1$ and $K\alpha_2$ doublet is observed as overlapping $K\alpha$ in many cases because of the limited energy resolution.

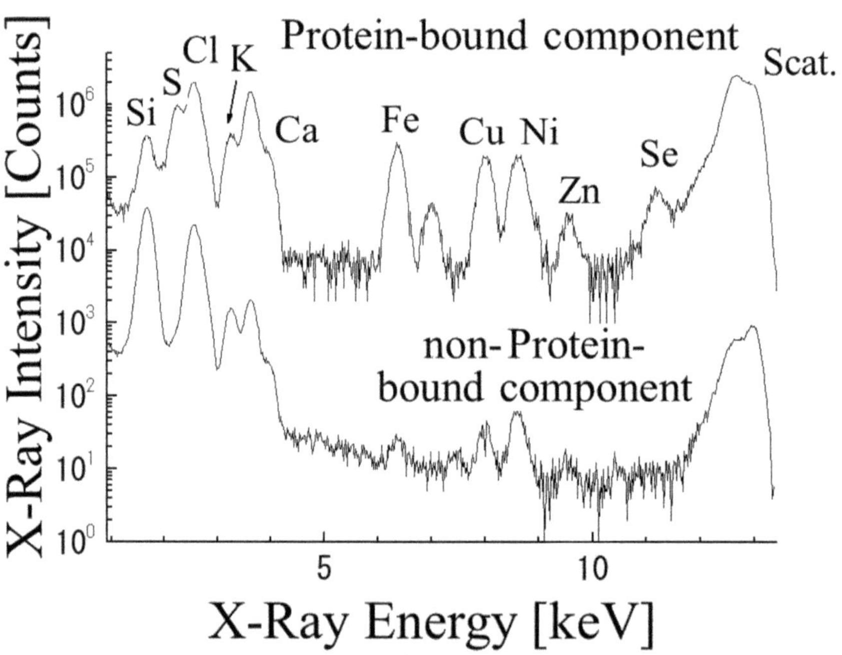

Figure 1.11. The x-ray fluorescence spectra of dried residue of 3 μl donor horse serum (ICN Biomedicals, USA) separated into protein-bound and non-protein-bound parts by centrifugation (10 000 G, 30 min, at 4 °C). Each peak can be attributed to a specific element, such as silicon (Si), sulfur (S), chlorine (Cl), potassium (K), calcium (Ca), iron (Fe), copper (Cu), nickel (Ni), zinc (Zn), or selenium (Se). A Si peak is from the silicon substrate. The data were collected at the Japanese synchrotron radiation facility, SPring-8. Reproduced from [84].

samples were prepared by centrifugation; one is the protein-bound component of the serum, and the other is the non-protein-bound component. Each peak in the x-ray fluorescence spectra corresponds to a chemical element contained in the sample. The peak position, i.e. the energy, characterizes the kind of chemical element, and the peak intensity gives the amount and/or the concentration. In this example, K and Ca are the major elements, and metals such as Fe, Cu, Ni, and Zn are minor or trace elements. Comparing the two spectra, the peaks of metals are clearly visible for the protein-bound component, but weak for the non-protein-bound component. The concentration of observed metals in the protein-bound component is in the order of 0.1 mM, while the concentration is much lower for the non-protein-bound component.

To perform quantification, the matrix effect must be considered. The matrix effect is the phenomenon in which: (a) the x-ray fluorescence intensity is affected by absorption in the sample itself, and (b) secondary and higher excitations of the x-ray fluorescence may contribute to the intensity. As in the case of other chemical analyses, in the calibration curve method, a series of standard samples is prepared containing a specific element at a known concentration in a very similar matrix. Nowadays, the reference-free approach, which does not require the preparation of a calibration curve, has become popular in x-ray fluorescence analysis. This method uses the theoretical x-ray intensity formula shown earlier and some physical constants known as 'x-ray fundamental parameters.' One of the successful software packages for such analysis, PIQUINT [85], which uses the database of x-ray fundamental parameters [50], was used for the exploration of the Martian surface with the PIXL instrument in the 2020 NASA Mars Rover Perseverance mission [86–88]. Because of the importance of obtaining accurate and reliable fundamental parameters, an international collaborative team is working on this [89]. The details of quantification in x-ray fluorescence spectroscopy are described in several textbooks [79–83].

X-ray fluorescence analysis has the following advantages:

1. It can analyze almost all elements from beryllium (Be, $Z = 4$) to californium (Cf, $Z = 98$). The K lines are measured for light elements, and L lines (or sometimes M lines) are measured for heavy elements.
2. It is non-destructive, so samples can be checked by other analytical methods even after x-ray measurements have taken place.
3. It works on solid, liquid, or gaseous samples.
4. Sample preparation is easy.
5. Measurements work in vacuum, air, or other gases.
6. It gives accurate, precise quantitative analysis.
7. Ultralow trace amounts of elements can be analyzed at levels as low as 10^{-16} g when highly brilliant x-ray sources and advanced x-ray spectrometers are employed [90].
8. The measurement time is short.

Modern x-ray fluorescence spectroscopy is easy to use, even for newcomers who are interested in getting results quickly. The analysis is almost automatic and is highly reliable in most cases. On the other hand, it is undesirable to use any analytical method as a black box. The following is therefore a minimum list of practical tips useful for energy-dispersive x-ray fluorescence spectroscopy using silicon drift detectors and/or other silicon-based semiconductor detectors.

The basics of practical energy-dispersive x-ray fluorescence analysis

1. Choose a primary x-ray energy higher than the absorption edge of the chemical element of interest. For example, to analyze molybdenum (Mo, $Z = 42$), set the x-ray tube voltage somewhat higher than 20 kV, as the K-absorption edge of molybdenum is 20.0 keV. In the case of synchrotron experiments, set the monochromator to use an energy somewhat higher than 20.0 keV.
2. Choose a suitable anode (such as Cr, Cu, W, Mo, Rh, Ag, etc.), a primary filter (material, thickness), and a tube voltage by considering the background spectral shape and the absorption edges of the chemical elements contained in the sample. Controlling the spectral shape of the primary x-rays is important because this shape overlaps as a background in the data. Therefore, it is important to remove the background in the energy region where signal peaks are expected. When the tube voltage is raised, the primary x-ray spectra move to higher energies. The use of a strong absorbing filter cleans up the lower-energy part of the primary x-ray spectrum. In the case of monochromatic excitation, such background is usually very low.
3. Peaks from neighboring elements may overlap. For example, the Ca Kα peak (3.69 keV) overlaps with that of K Kβ (3.59 keV). The Fe Kα peak (6.40 keV) overlaps with that of Mn Kβ (6.49 keV). The Co Kα peak (6.93 keV) overlaps with that of Fe Kβ (7.06 keV).
4. The peaks of light elements' K lines and heavy elements' L lines may overlap. For example, the peak of Ti Kα (4.51 keV) overlaps with that of Ba Lα (4.47 keV). The peak of Cu Kα (8.04 keV) overlaps with that of Ta Lα (8.15 keV). And the peak of Kα (10.54 keV) overlaps with that of Pb Lα (10.55 keV).
5. An Ar Kα (2.96 keV) peak is observed if the measurement is made in air. This peak is produced by Ar contained in air.
6. Low-energy x-ray peaks may not appear due to absorption by the detector window and the atmosphere. As Be windows are quite often used, energies lower than 1 keV fall outside the measurement range. For such soft x-rays, use vacuum and an ultrathin membrane.
7. Escape peaks can appear from interactions inside the detector. Such peaks appear at the lower-energy side of some strong signal peaks. In the case of silicon-based detectors, the energy difference between the escape peak and the signal peak is equal to the energy of Si Kα emission (1.74 keV).
8. Summed peaks commonly appear at high count rates. When Fe Kα emission (6.40 keV) is strong, a 12.8 keV peak may appear because the signal processor accidentally handles two incoming photons as a single photon.
9. X-ray fluorescence is absorbed by the sample itself, leading to a dependence of the fluorescent intensity on the concentration of other elements. On the other hand, secondary and higher excitations of x-ray fluorescence happen

depending on the concentrations of other elements. Those are called absorption effects and enhancement effects, respectively. Quantitative analysis considering such matrix effects is usually done via the calibration curve method using standard samples. Another promising approach is reference-free analysis using the theoretical formula and physical fundamental x-ray parameters. Modern commercially available systems handle both.

1.6 Nature is complex and inhomogeneous

As already discussed, x-ray fluorescence spectroscopy is an important tool for seeing nature, in view of the colors of the chemical elements. On the other hand, nature exhibits complexity, inhomogeneity, and nonuniformity. Figure 1.12 shows a Liesegang pattern, which is a self-organized concentric chemical ring and band pattern formed by simply pipetting some chemicals, such as silver nitrate ($AgNO_3$), onto a medium, such as ultrathin gelatin containing potassium dichromate ($K_2Cr_2O_7$) [91]. Before pipetting, chromium is uniformly distributed in the media, but after pipetting, chromium precipitates only in the rings by forming silver chromate. There is no chromium between rings. Any analysis of such a system is clearly limited if it neglects such patterns and gives the average of the whole area. Figure 1.13 shows a typical electrochemical deposit produced by a mixed aqueous solution of copper and zinc in a thin-layered circular cell [92]. The electrodeposit varies and exhibits five different patterns, i.e. dendritic, fine hair-like, bent, ramified,

Figure 1.12. Liesegang rings in ultrathin films. An optical photomicrograph of silver dichromate precipitation patterns formed at 5.0 ± 1.0 °C in ultrathin gelatin films with thicknesses of 65 nm. Reprinted with permission from [91]. Copyright (2016) American Chemical Society.

Figure 1.13. An optical photomicrograph of sparse electrodeposits of bent branches grown from a mixed electrolyte of Cu and Zn ions. A circular electrolysis cell was used with a copper ring anode (with an internal diameter of 16 mm) and a copper wire cathode (0.7 mm in dia.). The electrolyte (360 mM $ZnSO_4$ and 72 mM $CuSO_4$) was covered by 6 μm thick Mylar film and formed a layer approximately 250 μm thick inside the ring. Reprinted from [92], Copyright (2004), with permission from Elsevier.

and dense branching morphologies, depending on the ratio of copper and zinc concentrations in the electrolyte.

Self-organizing processes and diffusion–reaction systems spontaneously generate ordered and disordered structures in nature [93–95]. Phenomena such as crystal growth, convection patterns, and flocking behaviors are examples. Nonlinear reactions and interactions are also important. They can lead to feedback loops, sensitivity to initial conditions, chaos, fractals, and other complex behaviors [96, 97]. According to the studies of Ilya Prigogine (1917–2003), dissipative systems exchange energy and matter with their environment and can self-organize into new dynamic structures [98].

In addition, there are many competing interactions in nature at multiple scales [99]. For example, interactions that favor order vs. disorder, segregation vs. mixing, cooperation vs. competition, etc. The competition between these opposing interactions generates complexity. It is known that new properties and behaviors may sometimes emerge from the collective interactions between many simple components in a system, though it is often hard to predict from the properties of the individual components [100]. The systems often evolve and change over time, and then new structures and patterns emerge. Such dynamics and evolution over time are sources of complexity [101]. Furthermore, randomness, chance, and contingency may lead to complexity [102]. Even almost identical starting conditions may lead to very different outcomes, depending on chance events. This contributes to nonuniformity.

In summary, nature is far from mathematically perfect or simple because of self-organization, nonlinearity, the dissipation of energy, competing interactions, emergence, dynamics and evolution, and contingency.

To see nature, therefore, some spatial resolving power is crucial. For this, imaging capability will be key for scientific tools. As x-ray fluorescence spectroscopy is a way of determining the average chemical composition, the addition of an imaging capability is extremely desirable. X-ray color imaging is a solution to this demand to see nature's complexity, inhomogeneity, and nonuniformity.

References

[1] Fuller T 1732 *Introductio ad prudentiam: Part II Gnomologia* (London: Barker and Bettesworth Hitch)

[2] Kausche G A, Pfankuch E and Ruska H 1939 Die Sichtbarmachung von pflanzlichem virus im Übermikroskop *Naturwissenschaften* **27** 292–9 https://doi.org/10.1007/BF01493353

[3] 2022 Golden eye: A new space telescope makes a spectacular debut after a troubled gestation *Science* **378** 1160–1 https://doi.org/10.1126/science.adg2798

[4] Materion O-30-H Grade Beryllium https://materion.com/-/media/files/beryllium/specsheets/o-30-h.pdf

[5] Images from James Webb Space Telescope. https://webbtelescope.org/resource-gallery/images

[6] https://iupac.org/what-we-do/periodic-table-of-elements/

[7] Scerri E 2019 *The Periodic Table: Its Story and Its Significance* 2nd edn (Oxford: Oxford University Press)

[8] Mendeleev D 1869 System of elements according to their atomic weights and chemical functions *J. Prak. Chem.* **106** 251
Relationship of properties of the elements to their atomic weights *J. Russ. Chem. Soc. (in Russian)* **1** 60–77
1869 On the relations of properties of the elements to their atomic weights *Z. Chem.* **12** 405–6

[9] Mendeleev D 1870–1872 The natural system of the elements and its application to indication of properties of unknown elements (1870); Periodic regularity of the chemical elements (1871). Both were used as textbooks in his university class, and also were republished in *J. Russ. Phys.-Chem. Soc.* **3** 25–56 (in Russian) (1871) and
Annal. Chem. Pharm. Suppl. **7–8** 144–229 (in German)

[10] Conrad Roentgen W December 28, 1895 On a new kind of rays (in German) in the Proc. of the Würzburg Physico-Medical Society https://de.wikisource.org/wiki/Ueber_eine_neue_Art_von_Strahlen_(Vorl%C3%A4ufige_Mittheilung)

[11] Friedrich W, Knipping P and von Laue M 1912 Interference phenomena in x-rays *Sitzungsber. Bayerische Akad. Wiss* **42** 303–22 (in German)
von Laue MA quantitative test of the theory of interference phenomena in x-rays *Sitzungsber. Bayerische Akad. Wiss.* **42** 363–73 (in German)

[12] Bragg W H 1912 X-rays and crystals *Nature* **90** 219 https://doi.org/10.1038/090219a0

[13] Sagnac G 1901 Röntgen x-ray propagation (in French) *Ann. Chim. Phys. Paris* **22** 394–432

[14] Sagnac G 1901 Secondary rays derived from Röntgen rays (in French) *Ann. Chim. Phys. Paris* **22** 493–563

[15] Sagnac G 1901 Relation of x-rays and their secondary rays to matter and electricity (in French) *Ann. Chim. Phys.* **23** 145–98

[16] Barkla C G 1903 Secondary radiation from gases subject to x-rays *Phil. Mag. Ser.* **6** 685–98 https://doi.org/10.1080/14786440309462976

[17] Barkla C G 1903 Energy of secondary Röntgen radiation *Proc. Phys. Soc. London* **19** 185–204 https://doi.org/10.1088/1478-137814/19/1/319

[18] Barkla C G 1904 Polarisation in Röntgen rays *Nature* **69** 463 https://doi.org/10.1038/069463a0

[19] Barkla C G 1904 Energy of secondary Röntgen radiation *Phil. Mag. Ser.* **6** 543–60 https://doi.org/10.1080/14786440409463147

[20] Barkla C G 1905 Polarised Rontgen radiation *Proc. R. Soc. Lond.* **A74** 474–5 https://doi.org/10.1098/rspl.1904.0142

[21] Barkla C G 1905 Polarised Rontgen radiation *Phil. Trans. R. Soc.* **A204** 467–79 https://doi.org/10.1098/rsta.1905.0013

[22] Barkla C G 1905 Secondary Röntgen radiation *Nature* **71** 440 https://doi.org/10.1038/071440a0

[23] Barkla C G 1906 Secondary Rontgen rays and atomic weight *Nature* **73** 365 https://doi.org/10.1038/073365c0

[24] Barkla C G 1907 The nature of x-rays *Nature.* **76** 661–2 https://doi.org/10.1038/076661c0

[25] Barkla C G 1908 The nature of x-rays *Nature.* **78** 665 https://doi.org/10.1038/078665a0

[26] Barkla C G and Sadler C A 1909 The absorption of Röntgen rays *Phil. Mag. Ser.* **6** 739–60 https://doi.org/10.1080/14786440508636650

[27] Barkla C G 1909 Ionisation by Röntgen rays *Nature.* **80** 187 https://doi.org/10.1038/080187a0

[28] Barkla C G and Nicol J 1910 X-ray spectra *Nature.* **84** 139 https://doi.org/10.1038/084139a0

[29] Barkla C G and Nicol J 1911 Homogeneous fluorescent x-radiations of a second series *Proc. Phys. Soc. London* **24** 9 https://doi.org/10.1088/1478-7814/24/1/302

[30] Barkla C G 1911 Note on the energy of scattered x-radiation *Phil. Mag. Ser.* **6** 648–52 https://doi.org/10.1080/14786440508637077

[31] Barkla C G 1911 The spectra of the fluorescent Röntgen radiations *Phil. Mag. Ser.* **6** 396–412 https://doi.org/10.1080/14786440908637137

[32] Moseley H G J and Darwin C G 1913 The reflection of the x-rays *Nature.* **90** 594 https://doi.org/10.1038/090594a0

[33] Moseley H and Darwin C G 1913 The reflexion of the x-rays *Phil. Mag. Ser.* **6** 210–32 https://doi.org/10.1080/14786441308634968

[34] Moseley H G J 1913 The high-frequency spectra of the elements *Phil. Mag. Ser.* **6** 1024–34 https://doi.org/10.1080/14786441308635052

[35] Moseley H G J 1914 The high-frequency spectra of the elements, Part II *Phil. Mag. Ser.* **6** 703–13 https://doi.org/10.1080/14786440408635141

[36] Schoonjans T, Brunetti A, Golosio B, del Rio M S, Solé V A, Ferrero C and Vincze L 2011 The xraylib library for x-ray–matter interactions. Recent developments *Spectrochim. Acta.* **B66** 776–84 https://doi.org/10.1016/j.sab.2011.09.011 https://github.com/tschoonj/xraylib

[37] Siegbahn M 1916 On the high-frequency spectra (L-series) of the elements tantalum-uranium *Phil. Mag. Ser.* **6** 39–49 https://doi.org/10.1080/14786441608635542

[38] Siegbahn M 1916 On an x-ray vacuum spectrograph *Phil. Mag. Ser.* **6** 494–6 https://doi.org/10.1080/14786441608635594

[39] Siegbahn M 1919 Precision-measurements in the x-ray spectra *Phil. Mag. Ser.* **6** 601–12 https://doi.org/10.1080/14786440608635923

[40] Siegbahn M 1919 Precision-measurements in the x-ray spectra. Part II *Phil. Mag. Ser.* **6** 639–46 https://doi.org/10.1080/14786441108635992

[41] Siegbahn M 1919 Precision-measurements in the x-ray spectra. Part III *Phil. Mag. Ser.* **6** 647–51 https://doi.org/10.1080/14786441108635993

[42] Coster D and Hevesy G 1923 On the missing element of atomic number 72 *Nature.* **111** 79 https://doi.org/10.1038/111079a0

[43] Noddack W, Tacke I and Berg O 1925 Die Ekamangane *Naturwissenschaften* **13** 567–74 https://doi.org/10.1007/BF01558746

[44] Chapman K 2022 Masataka Ogawa and the search for nipponium *Chem. World* June 27 https://www.chemistryworld.com/opinion/masataka-ogawa-and-the-search-for-nipponium/4015784.article

[45] Poliakoff M Nipponium - The Element That Wasn't - Periodic Table of Videos https://youtu.be/kfYCvbdX9Ck?si=mgmbyI4HWnpkAq_g

[46] Perrier C and Segrè E 1947 Technetium: the element of atomic number 43 *Nature* **159** 24 https://doi.org/10.1038/159024a0

[47] Perey M 1939 L'élément 87: AcK, dérivé de l'actinium (in French) *J. Phys. Radium* **10** 435–8 https://doi.org/10.1051/jphysrad:019390010010043500

[48] Corson D R, MacKenzie K R and Segrè E 1940 Artificially radioactive element 85 *Phys. Rev.* **58** 672 https://doi.org/10.1103/PhysRev.58.672

[49] Marinsky J A, Glendenin L E and Coryell C D 1947 The chemical identification of radioisotopes of neodymium and of element 61 *J. Am. Chem. Soc.* **69** 2781–5 https://doi.org/10.1021/ja01203a059

[50] Elam W T, Ravel B D and Sieber J R 2002 A new atomic database for x-ray spectroscopic calculations *Radiat. Phys. Chem.* **63** 121–8 https://doi.org/10.1016/S0969-806X(01)00227-4 https://github.com/xraypy/XrayDB

[51] Giliam E and Heal H T 1952 Some problems in the analysis of steels by x-ray fluorescence *Br. J. Appl. Phys.* **3** 353–8 https://doi.org/10.1088/0508-3443/3/11/304

[52] Sherman J 1955 The theoretical derivation of fluorescent x-ray intensities from mixtures *Spectrochim. Acta.* **7** 283–306 https://doi.org/10.1016/0371-1951(55)80041-0

[53] Sherman J 1958 A theoretical derivation of the composition of mixable specimens from fluorescent x-ray intensities *Adv. X-ray Anal.* **1** 231–50 https://doi.org/10.1154/S0376030800000197

[54] Sherman J 1959 Simplification of a formula in the correction of fluorescent x-ray intensities from mixtures *Spectrochim. Acta* **11** 466–70 https://doi.org/10.1016/S0371-1951(59)80341-6

[55] Shiraiwa T and Fujino N 1966 Theoretical calculation of fluorescent x-ray intensities in fluorescent x-ray spectrochemical analysis *Jpn. J. Appl. Phys.* **5** 886–99 https://doi.org/10.1143/JJAP.5.886

[56] Shiraiwa T and Fujino N 1967 Theoretical calculation of fluorescent x-ray intensities of nickel-iron-chromium ternary alloy *Bull. Chem. Soc. Jpn.* **40** 2289–96 https://doi.org/10.1246/bcsj.40.2289

[57] Criss J W and Birks L S 1968 Calculation methods forfluorescent x-ray spectrometry—empirical coefficients vs. fundamental parameters *Anal. Chem.* **40** 1080–6 https://doi.org/10.1021/ac60263a023

[58] Laguitton D and Parrish W 1977 Simultaneous determination of composition and mass thickness of thin films by quantitative x-ray fluorescence analysis *Anal. Chem.* **49** 1152–6 https://doi.org/10.1021/ac50016a023

[59] Criss J W, Birks L S and Gilfrich J V 1978 Versatile x-ray analysis program combining fundamental parameters and empirical coefficients *Anal. Chem.* **50** 33–7 https://doi.org/10.1021/ac50023a013

[60] Knoll G F 2010 *Radiation Detection and Measurement* 4th edn (New York: Wiley)

[61] Tsoulfanidis N 2015 *Measurement and Detection of Radiation (English Edition)* 4th edn (Boca Raton, FL: CRC Press) https://doi.org/10.1201/b18203

[62] Tait W H 1980 *Radiation Detection* (Oxford: Butterworth-Heinemann)

[63] Gerhard L 2007 *Semiconductor Radiation Detectors—Device Physics"* (Berlin: Springer) https://doi.org/10.1007/978-3-540-71679-2

[64] Cerrito L 2017 *Radiation and Detectors: Introduction to the Physics of Radiation and Detection Devices* (Berlin: Springer) https://doi.org/10.1007/978-3-319-53181-6

[65] Sakai E 1982 Present status of room temperature semiconductor detectors *Nucl Instrum. Methods* **196** 121–30 https://doi.org/10.1016/0029-554X(82)90626-7

[66] van Heerden P J 1945 The crystal counter—a new instrument in nuclear physics *PhD Thesis* (Amsterdam: N. V. Noord-Hollandsche Uitgevers Maatschappij)

[67] McKay K G 1949 A germanium counter *Phys. Rev.* **76** 1537 https://doi.org/10.1103/PhysRev.76.1537

[68] McKay K G 1951 Electron-hole production in germanium by alpha-particles *Phys. Rev.* **84** 829 https://doi.org/10.1103/PhysRev.84.829

[69] Pell E M 1960 Ion drift in an n-*p* junction *J. Appl. Phys.* **31** 291–302 https://doi.org/10.1063/1.1735561

[70] Klein C A 1968 Bandgap dependence and related features of radiation ionization energies in semiconductors *J. Appl. Phys.* **39** 2029–38 https://doi.org/10.1063/1.1656484

[71] Fano U 1947 Ionization yield of radiations. II. The fluctuations of the number of ions *Phys. Rev.* **72** 26 https://doi.org/10.1103/PhysRev.72.26

[72] Sakai E 1968 Slow pulses from a planer Ge(Li) detector *Appl. Phys. Lett.* **12** 269–71 https://doi.org/10.1063/1.1651987

[73] Ahamad I and Wagner F 1974 A simple cooled Si(Li) electron spectrometer *Nucl. Instrum. Methods* **116** 465–9 https://doi.org/10.1016/0029-554X(74)90828-3

[74] Gatti E and Rehak P 1984 Semiconductor drift chamber—an application of a novel charge transport scheme *Nucl. Instrum. Methods* **225** 608–14 https://doi.org/10.1016/0167-5087(84)90113-3

[75] Gatti E, Rehak P and Watson J T 1984 Silicon drift chambers—first results and optimum processing of signals *Nucl. Instrum. Methods* **226** 129–41 https://doi.org/10.1016/0168-9002(84)90181-5

[76] Kemmer J and Lutz G 1987 New detector concepts *Nucl. Instrum. Methods* **253** 365–77 https://doi.org/10.1016/0168-9002(87)90518-3

[77] Kemmer J, Lutz G, Belau E, Prechtel U and Welser W 1987 Low capacity drift diode *Nucl. Instrum. Methods* **253** 378–81 https://doi.org/10.1016/0168-9002(87)90519-5

[78] Lechner P, Pahlke A and Soltau H 2004 Novel high-resolution silicon drift detectors *X-ray Spectrom.* **33** 256–31 https://doi.org/10.1002/xrs.717
[79] Jenkins R 1995 *Quantitative X-Ray Spectrometry* 2nd edn (Boca Raton, FL: CRC Press) https://doi.org/10.1201/9781482273380
[80] Van Grieken R and Markowicz A 2001 *Handbook of X-Ray Spectrometry* 2nd edn (Boca Raton, FL: CRC Press) https://doi.org/10.1201/9780203908709
[81] Beckhoff E B, Kanngieser B, Langhoff N, Wedell R and Wolff H 2007 *Handbook of Practical X-Ray Fluorescence Analysis* (Berlin: Springer) https://doi.org/10.1007/978-3-540-36722-2
[82] Willis J, Turner K and Pritchard G 2011 XRF in the workplace: a guide to practical XRF spectrometry (Chipping Norton, NSW: Malvern PANalytical)
[83] Sakurai K (ed) 2019 *Introduction to Reference-Free X-ray Fluorescence Analysis (in Japanese)* (Tokyo: Kodan-sha) https://www.kspub.co.jp/book/detail/5135983.html
[84] Sakurai K, Eba H and Goto S 1999 Grazing incidence x-ray fluorescence and scattering experiments at BL-39XU, SPring-8 *Jpn. J. Appl. Phys. Suppl.* **38–1** 332–5 https://doi.org/10.7567/JJAPS.38S1.332
[85] Heirwegh C M, Elam W T, O'Neil L P, Sinclair K P and Das A 2022 The focused beam x-ray fluorescence elemental quantification software package PIQUANT *Spectrochim. Acta Part* B **19** 1020 https://doi.org/10.1016/j.sab.2022.106520
[86] Heirwegh C M, Elam W T, Flannery D T and Allwood A C 2018 An empirical derivation of the x-ray optic transmission profile used in calibrating the Planetary Instrument for X-ray Lithochemistry (PIXL) for Mars 2020 *Powder Diffr.* **33** 162–5 https://doi.org/10.1017/S0885715618000416
[87] Allwood A C, Wade L A, Foote M C, Elam W T *et al* 2020 PIXL: planetary instrument for x-ray lithochemistry *Space Sci. Rev.* **216** 134 https://doi.org/10.1007/s11214-020-00767-7
[88] Heirwegh C 2023 Methods and reference materials used to calibrate PIXL, the Mars 2020 *in situ* XRF spectrometer *Microsc. Microanal.* **29** 235–6 https://doi.org/10.1093/micmic/ozad067.105
[89] FP initiative—international initiative on x-ray fundamental parameters. https://www.exsa.hu/?inh=635
[90] Sakurai K, Eba H, Inoue K and Yagi N 2002 Wavelength-dispersive total-reflection x-ray fluorescence with an efficient Johansson spectrometer and an undulator x-ray source: detection of 10–16 g-level trace metals *Anal. Chem.* **74** 4532–5 https://doi.org/10.1021/ac025720y
[91] Jinxing Jiang and Sakurai K 2016 Formation of ultrathin liesegang patterns *Langmuir.* **32** 9126–34 https://doi.org/10.1021/acs.langmuir.6b02148
[92] Eba H and Sakurai K 2004 Pattern transition in Cu-Zn binary electrochemical deposition *J. Electroanal. Chem.* **571** 149–58 https://doi.org/10.1016/j.jelechem.2004.05.024
[93] Kauffman S A 1993 *The Origins of Order: Self-Organization and Selection in Evolution* (Oxford: Oxford University Press) https://doi.org/10.1093/oso/9780195079517.001.0001
[94] Grzybowski B A 2009 *Chemistry in Motion: Reaction-Diffusion Systems for Micro- and Nanotechnology* (New York: Wiley) https://doi.org/10.1002/9780470741627
[95] ben-Avraham D and Havlin S 2000 *Diffusion and Reactions in Fractals and Disordered Systems* (Cambridge: Cambridge University Press) https://doi.org/10.1017/CBO9780511605826
[96] Berge P, Pomeau Y and Vidal C 1987 *Order within Chaos* 1st edn (New York: Wiley-VCH)

[97] Strogatz S H 2015 *Nonlinear Dynamics and Chaos: With Applications to Physics, Biology Chemistry, and Engineering* 2nd edn (Boulder, CO: Westview Press) https://doi.org/10.1201/9780429492563

[98] Nicolis G and Prigogine I 1977 *Self-Organization in Nonequilibrium Systems: From Dissipative Structures to Order through Fluctuations* (New York: Wiley) https://doi.org/10.1002/bbpc.197800155

[99] Vedmedenko E 2007 *Competing Interactions and Pattern Formation in Nanoworld* (New York: Wiley-VCH) https://doi.org/10.1002/9783527610501

[100] Luisi P L 2016 *The Emergence of Life: From Chemical Origins to Synthetic Biology* (Cambridge: Cambridge University Press) https://doi.org/10.1017/CBO9781316135990

[101] Pismen L 2023 *Patterns and Interfaces in Dissipative Dynamics* 2nd edn (Berlin: Springer) https://doi.org/10.1007/978-3-031-29579-9

[102] Gray P, Nicolis G, Baras F, Borckmans P and Scott S K (ed) 1992 *Spatial Inhomogeneities and Transient Behaviour in Chemical Kinetics* (New York: Wiley) https://www.wiley.com/en-us/Spatial+Inhomogeneities+and+Transient+Behaviour+in+Chemical+Kinetics-p-9780471934974

IOP Publishing

X-ray Color Imaging
Static and dynamic x-ray fluorescence for chemical element identification
Kenji Sakurai and Wenyang Zhao

Chapter 2

What does x-ray color imaging look like?

2.1 The scanning and projection types of x-ray fluorescence imaging

The observation of the complexity, inhomogeneity, and nonuniformity of nature is greatly facilitated by the use of x-rays, which possess the capability of discerning chemical elements as color within the x-ray wavelength region. In particular, when coupled with an imaging capability, the technique, namely x-ray color imaging, exerts a pronounced influence. Typically, x-ray color imaging is x-ray fluorescence (XRF) imaging, which is based on the energy-dispersive detection of XRF spectra from each chemical element present in the sample. Figure 2.1 illustrates the recent methodological evolution of XRF analysis. Since the mid-20th century, energy-dispersive XRF has been extensively utilized for practical analysis to determine the chemical composition of various unknown samples.

Importantly, XRF has become capable of providing an imaging capability through the use of a micro primary x-ray beam and an *x*–*y* positional scanner for the sample. This is commonly referred to as scanning-type imaging. During scanning, XRF spectra are measured and recorded for each point. By analyzing the XRF spectra, the chemical composition for each point can be determined. By integrating such information about the chemical composition for all points, maps for each chemical element contained in the sample can be generated. Scanning-type imaging has been employed worldwide since the late 20th century, especially at synchrotron light sources [1–4]. With the incorporation of suitable x-ray optics, it is also used in independent laboratories [5–7]. In addition, commercial micro-XRF machines with a scanning-type imaging capability are now available.

Toward the end of the 20th century and the beginning of the 21st century, a new trend of projection-type imaging emerged. Its operation is similar to the capture of ordinary camera photos, where images are projected onto a camera sensor through a lens system. The new method allows the simultaneous acquisition of chemical composition for all points on the sample [8]. In this way, projection-type imaging eliminates the need for point-by-point scanning. It is also sometimes referred to as

doi:10.1088/978-0-7503-3215-6ch2

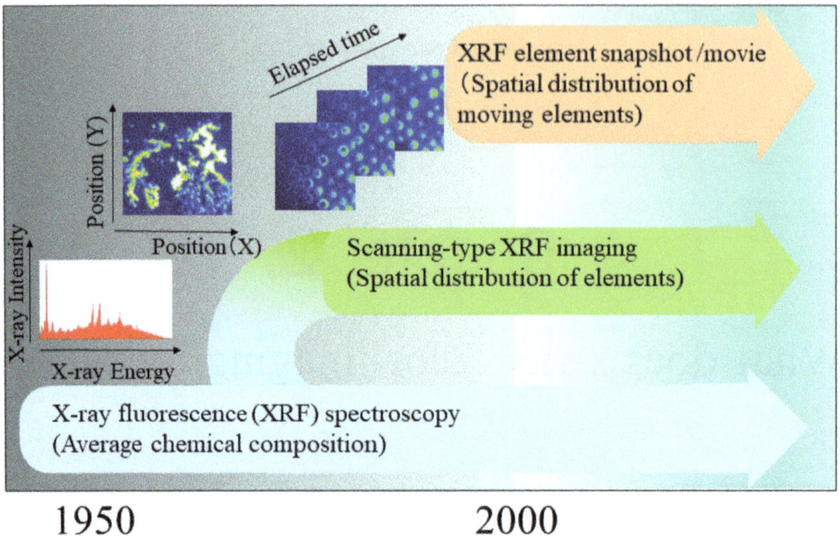

Figure 2.1. The methodological development of x-ray fluorescence analysis. In the mid-20th century, the advent of semiconductor detectors made energy-dispersive x-ray fluorescence measurement possible, allowing simultaneous multielement analysis. Later, the combination of x-ray focusing optics and *x–y* positional scanning of the sample made it possible to image spatially distributed chemical elements. In the late 20th century, another breakthrough came with the introduction of 2D semiconductor detectors, which allowed x-ray color imaging without the need for *x–y* scanning.

Table 2.1. A comparison of methods for visualizing the colors of chemical elements using x-rays. There are two types of x-ray color imaging methods: scanning-type imaging and projection-type imaging. In this book, the focus is primarily on the newer trend of projection-type imaging.

	XRF analysis	Scanning-type XRF imaging	Projection-type XRF imaging
Available information	Average chemical composition	Spatial distribution of each element (stationary)	Spatial distribution of each element (stationary and moving)
Typical spatial resolution	—	At synchrotron: 0.02–2 μm. At independent labs: 5–100 μm.	10–50 μm
Typical pixel number	—	100 × 100	1000×1000 or more
Typical measurement time	A few seconds to 10 min	A few hours or more	100 ms to a few minutes

full-field imaging because the measurement simultaneously covers the entire field of view, rather than sequentially measuring selected microregions.

Table 2.1 summarizes the performance of XRF analysis, scanning-type XRF imaging, and projection-type XRF imaging. It is worth noting that sometimes, in order to obtain high-quality XRF images with strong XRF signals, the measurement time for projection-type imaging can be extended to a few hours. However, when a large number of pixels are required, projection-type imaging is undoubtedly significantly faster than scanning-type imaging. For cases where energy resolution is not always important (if strict separation between chemical elements contained in the sample is not required), the measurement time for projection-type imaging can be as short as 10–100 milliseconds or even less, enabling ultrafast movies. In situations where good energy resolution is required, a set of XRF images of different elements can also be obtained within a reasonable time, such as tens of seconds to a few minutes. This short measurement time allows for *in operando* movie imaging of changing samples, in which the spatial distribution of chemical elements continuously evolves.

In practice, projection-type imaging has long been limited by two main factors. The first limitation is that, unlike visible light, there are no efficient convex or concave lenses for x-rays [9]. Even now, advanced x-ray optical components used as 'lenses' for projecting x-ray images, such as reflection mirrors and polycapillary optics, are still inefficient but expensive. The achievable spatial resolution is limited to a few microns. On the other hand, simple x-ray optical components, such as the micro pinhole, are frequently used for projecting x-ray images, albeit with reduced imaging efficiency and spatial resolution.

The second limitation concerns color x-ray cameras. In the late 20th century, the advent of 2D semiconductor detectors made projection-type XRF imaging possible. However, most of these detectors could only capture black-and-white images of x-ray intensity maps. The information regarding x-ray photon energy, i.e. x-ray wavelength or color, was lost, making it impossible to identify elements or separate different elements. Over the last two decades, significant efforts have been made to develop advanced color x-ray cameras. Meanwhile, an alternative approach is becoming popular, which is the unconventional use of commercially available visible-light digital cameras. In this process, the camera sensors are directly exposed to x-rays. Thousands of camera snapshots are taken with short exposure times, resulting in the recording of only a few x-ray photons in each snapshot. These x-ray photons are likely to be scattered at different positions in the image, allowing their energies to be individually retrieved. When operated in this special 'single-photon-counting' mode, high-quality visible-light camera sensors, either charge-coupled devices (CCDs) [10] or complementary metal–oxide–semiconductor (CMOS) sensors [11], can function as color x-ray cameras. When combined with a simple x-ray optical component such as a micro pinhole, projection-type XRF imaging can be performed in a cost-effective manner which is easily accessible for independent laboratories, schools, and factories. Detailed explanations of x-ray optical components and color x-ray cameras are provided in chapter 3.

In comparison to scanning-type XRF imaging, the greatest advantage of projection-type XRF imaging is its ability to perform movie imaging of samples undergoing reactions, in which the spatial distribution of chemical elements continuously evolves. Rapid motion and the gradual diffusion of chemical elements are important topics in the fields of chemistry and materials science. These subjects encompass fascinating chemical reactions such as the Belousov–Zhabotinsky reaction, chemical gardens, and Liesegang rings, as well as practical engineering issues such as corrosion, electrodeposition, powder sintering, alloy welding, and diffusion at functional interfaces. For most issues, there is no other effective method for directly visualizing the motion of chemical elements. An indirect method is to record the changes in morphology and colors under optical microscopy. However, in terms of element analysis, XRF imaging is much more reliable than optical microscopy, as XRF signals originate from atoms, and the identification of chemical elements is not affected by their valences, concentrations, crystal structures, and coexistences. In addition, as an *in operando* technique, XRF movie imaging is applicable to samples in various states, including reactions at high temperatures or beneath a thin aqueous layer. Certainly, as XRF movie imaging is an emerging technique, its spatial and temporal resolutions are still limited at present.

For the purpose of mapping chemical elements, electron probe micro analysis (EPMA) is another well-known technique. In this technique, a micro electron beam is scanned across the sample to analyze secondary electrons, characteristic x-rays, Auger electrons, and other emissions. When used to analyze characteristic x-rays, this technique provides maps of the chemical elements present on a sample's surface; it is similar to scanning-type XRF imaging except that the probe consists of electrons rather than x-rays. The typical spatial resolution of a desktop EPMA is around 0.3–2 microns due to the fine focus of the primary electron beam. So far, EPMA has been widely applied in the popularization of scanning electron microscopy (SEM). Nevertheless, XRF imaging has several distinct advantages over EPMA. First, EPMA requires the sample to be tested in a vacuum environment, whereas XRF imaging does not. This has limited the application of EPMA. For example, wet samples cannot be brought into vacuum, and some samples are too large for a vacuum chamber. Second, EPMA is surface sensitive. Its probing depth is typically one or two microns. In contrast, the probing depth of XRF imaging is much larger. For example, if the target chemical element is iron, the probing depth in ore can exceed 100 microns, while in water, it can reach up to 1 mm. Third, insulating samples need to be coated with a thin conductive layer before EPMA takes place, whereas XRF imaging does not require any sample pretreatment. Fourth, a high-energy electron beam may cause radiation damage to the sample, and therefore caution should be exercised when using EPMA for irreplaceable, valuable, or fragile samples. In such cases, XRF imaging can be safer. So far, XRF imaging has been widely applied to the study of many valuable oil paintings.

The following sections of this chapter introduce the applications of projection-type XRF imaging and provide numerous examples from the fields of biology, mineralogy, industry, cultural heritage, and chemistry. All images were acquired using projection-type XRF imaging rather than scanning-type XRF imaging. The experimental information and the instruments utilized are briefly listed. Detailed explanations of the instruments can be found in chapter 3.

2.2 Biology

When Röntgen first discovered x-rays in 1895, he took a photo of the bones inside a hand, utilizing x-rays' high penetrative power and their differential absorption by materials with different densities and atomic numbers. Nowadays, x-ray radiography (x-ray transmission imaging) is still extensively utilized for examining the human body in hospitals. This is the most popular and well-known application of x-rays in biology.

However, the application of x-rays in biology is not limited to this. XRF imaging can be utilized to study the distribution of chemical elements in biological tissues [12, 13]. Generally, the photon energy of XRF analysis (<20 keV) is lower than that of x-ray radiography (~70 keV), and therefore the penetrative power is much weaker. However, the probing depth can still reach hundreds of microns to two or three millimeters, depending on the composition and density of the biological tissue. This probing depth is sufficient for analyzing the vast majority of sliced biological samples, as well as insects and small animals.

In the microscopic study of biological samples, staining is commonly used to highlight the accumulation of certain chemical substances, increase their contrast under the microscope, and distinguish different biological tissues. In neuroscience, calcium indicators, whether chemical or genetically encoded, are employed to trace the release and uptake of calcium ions, revealing the transmission of the accompanying neural signals. In contrast to these biological techniques, XRF signals originate within atoms and thus are inherently sensitive to the chemical elements, regardless of their physical or chemical states. Therefore, XRF imaging can work directly for analyses related to chemical elements, without the need for staining or chemical labeling.

Chemical elements play crucial roles in organisms. Carbon, hydrogen, oxygen, nitrogen, and phosphorous are the major elements that make up biological macromolecules. Free potassium, sodium, and chloride ions can maintain osmotic pressure and biological membrane potential. Some metal ions often bind to coenzyme groups to alter the conformation of enzymes and activate their catalytic functions. In plants, potassium and calcium are macronutrients. Chlorine, iron, manganese, zinc, copper, nickel, molybdenum, etc. are micronutrients. Conventionally, plants are burned to ash and then analyzed to determine the contents of these elements. However, using XRF imaging, it is possible to directly visualize the distribution of these elements in living plants and even monitor their transport. In animals, a common example is iron. It can bind to hemoglobin, which is responsible for transporting oxygen. Therefore, iron may accumulate in tissues that are rich in capillaries or related to the metabolism of red blood cells, such as the liver, spleen, and bone marrow.

Another meaningful application is the detection of trace-level pollution caused by toxic heavy metal elements in the environment. These toxic metal elements can accumulate in organisms, making their detection easier. Depending on the route of intake and metabolism, the organs where toxic metal elements accumulate may differ. For small insects, such accumulation can be directly visualized by XRF imaging, without the need for sample preparation. This is also significant for studying the impact of environmental pollution on human health. Typical toxic heavy metal elements include cadmium, mercury, arsenic, and lead.

In biological applications of XRF imaging, the most challenging issue is the limit of detection (LOD). Most chemical elements in biological tissues are present at very low levels. While imaging abundant elements such as chlorine, potassium, and calcium is relatively straightforward, imaging other chemical elements at trace levels may require LODs as low as 100 ppm or even 10 ppm. This requires careful

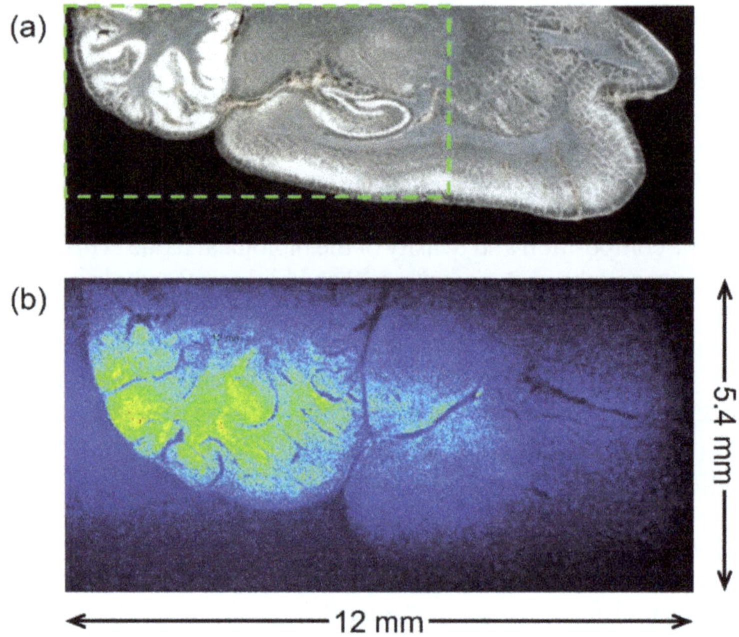

Figure 2.2. (a) A photomicrograph of a mouse brain slice. The dashed rectangle encloses the XRF imaging area. (b) XRF imaging of the slice. The bright part on the left is rich in iron. Sample courtesy of Professor I. Nakai from the University of Tokyo. Modified from figure 2 in reference. Adapted with permission from [14]. Copyright (2003) American Chemical Society.

instrument design for XRF imaging and will remain a challenging issue for a long time to come.

The first illustrative example is the use of XRF imaging in the study of the brain [14]. Figure 2.2(a) shows a microscopic image of an unstained mouse brain slice. The dashed rectangle encloses a viewing area of 12 mm × 5.4 mm. The XRF imaging experiment was performed at BL-4A, Photon Factory, KEK, Japan, using monochromatic primary x-rays from a normal bending magnet synchrotron source. The photon energy of the primary x-rays was set to 7.2 keV, just above the iron K-edge (7.112 keV). The detector was a conventional CCD camera (TC215 CCD, from Texas Instruments, 1000×1018 pixels, pixel size 12 μm, cooled to −30 °C). The x-ray optical component was a 1 mm thick collimator plate with capillaries 6 μm in diameter capable of projecting 1:1 magnification images. The spatial resolution achieved was 20 μm. The exposure time for the XRF image in figure 2.2(b) was 100 s.

Since the detector utilized was unable to resolve the x-ray energies, the intensity in the XRF image may come from the XRF signals of multiple elements that were excited by the 7.2 keV primary x-rays. However, during the experiment, we noticed that the left part of the XRF image abruptly became bright when the photon energy of the primary x-rays was increased to just above the iron K-edge. This indicated that the XRF signals were mainly iron Kα (6.403 keV) and iron Kβ (7.057 keV) and that the XRF image roughly represented a map of iron.

Let us compare the XRF image with the photomicrograph. In the photomicrograph, the white layer in the lower left corner corresponds to the granular layer of the cerebellum, which is the deeper layer of the cortex and contains numerous densely packed small cells. The iron-rich region revealed in the XRF image corresponds with the region beneath the granular layer. For many years, the accumulation of iron in brain has remained an interesting topic. Iron accumulation may occur in different parts of the brain, such as the basal ganglia or cerebellum, often accompanied by optic atrophy, Parkinson's disease, dystonia, psychiatric manifestations, and cognitive decline [15]. Some reports describe the use of magnetic resonance imaging (MRI) to characterize the accumulation of iron in the brain. However, as demonstrated in this example, XRF imaging may provide a new approach for characterizing fixed samples. The reader may observe a high degree of similarity between this XRF image and the map of iron obtained by ultrahigh-resolution MRI [16], while the corresponding neural mechanisms are still under discussion.

Another interesting biological example suitable for chemical element mapping by x-rays is the body of a bee. First, studying trace elements in bees is useful to gain a better understanding of how the environment affects their health and the quality of the honey they produce. Copper, chromium, zinc, manganese, and iron are commonly found in bee tissues, and they seem to originate from human activities and industries [17]. Studying the metal content in bees can serve as a crucial biomonitoring tool that allows us to track changes in the natural environment. According to one report [18], the concentration of elements in bee tissues were measured as follows: iron (249 $\mu g\ g^{-1}$), potassium (11,182 $\mu g\ g^{-1}$), manganese (59 $\mu g\ g^{-1}$), zinc (200 $\mu g\ g^{-1}$), and chromium (6.1 $\mu g\ g^{-1}$). These values reflect the typical levels of elemental concentrations. Among them, the concentration of chromium is of particular concern, as it may indicate environmental pollution.

Obviously, in addition to the average concentration of each chemical element in the bee body, it is important to map them. Figure 2.3 is an artistic representation of the typical distribution of the chemical elements potassium, calcium and iron in a bee body. It is known that potassium is almost uniformly distributed throughout the bee body, while iron and calcium are enriched at the wing attachment sites on the thorax and at the sting on the tail. It is possible to study further details of the distribution of each chemical element by the projection-type XRF imaging using a laboratory x-ray tube with a molybdenum anode and a conventional CCD camera (from Andor, 1024 × 1024 pixels, pixel size 13 μm, sensor thickness 40 μm, cooled to −85 °C, single-photon-counting mode to obtain an energy resolution of 150 eV at 5.9 keV) [19].

The third example is the XRF imaging of daphnia. Daphnia is a sentinel genus for environmental health protection and is currently widely used as one of the biological indicators to assess the level of water pollution. By observing the mortality of parent animals and the reproduction rate of daphnia in sampled water, researchers can determine the so-called 'no observed effect' concentrations of environmentally relevant substances [20, 21].

Figures 2.4 and 2.5 show XRF images of two daphnia [22]. One daphnia was cultured in the standard culture medium and the other was exposed to an additional 2.5 $mg\ l^{-1}$ of zinc and nickel for one day. The XRF imaging experiment was carried

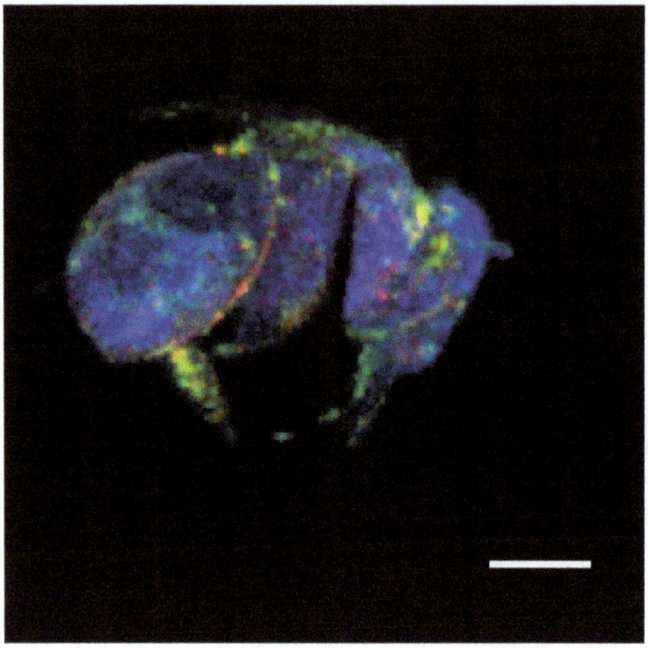

Figure 2.3. An artistic representation of the distribution of potassium (blue), calcium (green) and iron (red) in a typical bee body. The scale bar is 2 millimeters.

out at BAMline, BESSY, Germany utilizing monochromatic primary x-rays. The detector was a pnCCD (264×264 pixels, pixel size 48 μm, cooled to −25 °C) which was specially designed for x-ray color imaging; it had an energy resolution of 152 eV at 5.9 keV. The optical component was a micro pinhole (25 μm or 50 μm in diameter) manufactured in platinum covered with a tungsten layer; however, the imaging magnification via the pinhole was not reported. The experimental conditions were 90 min measurement with 13.5 keV primary x-rays for figure 2.4, 90 min measurement with 13.5 keV primary x-rays and 150 min measurement with 10 keV primary x-rays for the left-hand part of figure 2.5(a), and 240 min measurement with 10 keV primary x-rays for the right-hand part of figure 2.5(b). The energy of the primary x-rays was always higher than the K-edges of the chemical elements to be imaged, e.g. potassium (3.607 keV), calcium (4.038 keV), manganese (6.540 keV), iron (7.112 keV), nickel (8.333 keV) and zinc (9.659 keV).

In figure 2.4 we can see the maps of different elements in the daphnia cultured in the standard culture medium: calcium is abundantly present in the exoskeleton (a) and antennae (b). Iron is primarily detected in the thoracic appendages that carry gill-like tissues (c), the digestive diverticula (d), and the hindgut (e). Manganese is almost exclusively detected in the gut. Zinc and potassium exhibit comparable distribution patterns, with relatively higher levels in the midgut area (f) and embryos developing in the brood chamber (g). In addition, the compound eyes (h) show a high concentration of potassium.

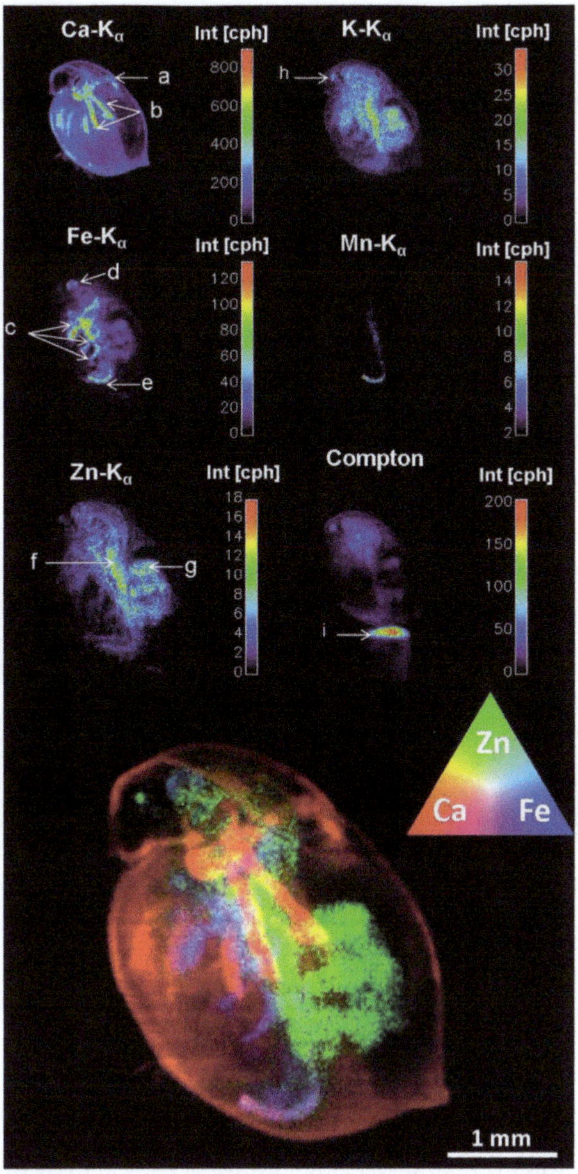

Figure 2.4. Elemental maps of calcium, potassium, iron, manganese, and zinc for a daphnia cultured in the standard culture medium, along with the distribution of Compton scattering in the measurement. The RGB color representations at the bottom of the image are red for calcium, green for zinc, and blue for iron, respectively. Adapted from figure 4 in reference. Reproduced from [22] with permission from the Royal Society of Chemistry.

The most interesting result is a comparison of the elemental maps of the daphnia cultured in the standard culture medium and the daphnia exposed to zinc/nickel, as shown in figure 2.5(a). Figure 2.5(b) shows the corresponding XRF spectra. In the

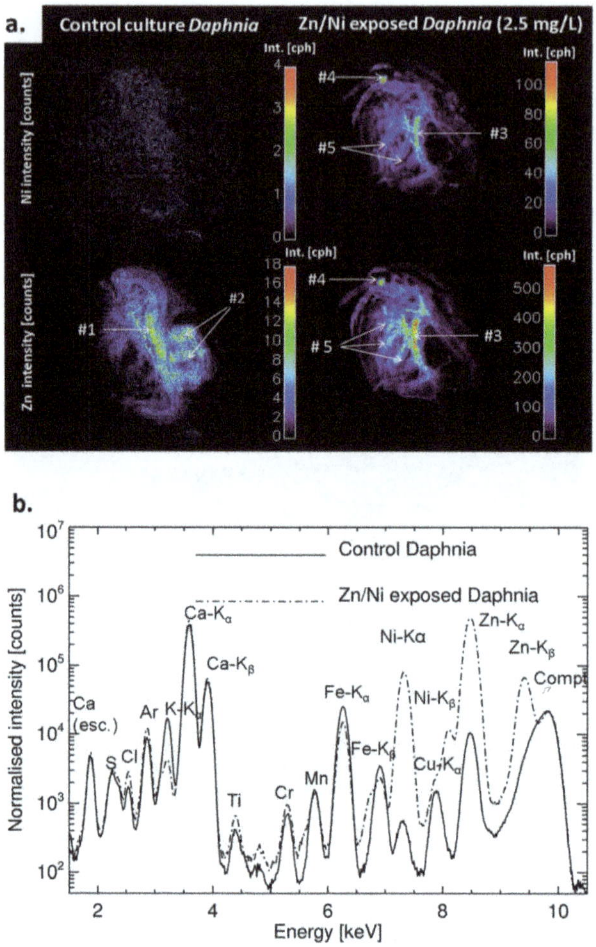

Figure 2.5. (a) Elemental maps of nickel and zinc in the control daphnia and daphnia exposed to nickel/zinc. (b) The corresponding summed XRF spectra. Adapted from figure 5 in reference. Reproduced from [22] with permission from the Royal Society of Chemistry.

daphnia cultured in the standard culture medium, the concentration of nickel is close to or below the limit of detection, resulting in a low-count, noisy distribution in the XRF image. The midgut region (#1) and embryos (#2) exhibit the highest levels of zinc. In the exposed daphnia, nickel and zinc have a similar distribution, seen in the regions of the midgut (#3), the compound eye (#4), and the region of the thoracic appendages (#5). These high levels of nickel and zinc can also be observed in the XRF spectra. The regions where nickel and zinc accumulate imply the presence of their metabolic pathways within the daphnia's body and show which of the daphnia's organs are susceptible to heavy metal contamination in the environment.

Due to the complex structure of biological organisms, accurately determining the location of chemical elements within their bodies through a 2D projection image can sometimes be challenging. In such cases, researchers can continuously rotate the sample, collect a series of XRF images taken at different angles, and reconstruct 3D maps of the elements. This process is similar to the CT scans used for human body examinations in hospitals, although there is a notable difference: in CT scans, each x-ray transmission image is a direct projection of human tissues from a point x-ray source. However, in 3D XRF imaging, each atom emitting fluorescence x-rays acts like a point source, and the collected XRF image is the projection of these sources through an x-ray optical component.

As an illustration, figure 2.6 shows the 3D XRF imaging of calcium and iron in a single foraminifer. Foraminifera are ancient protists that often possess a complex external shell primarily composed of calcium carbonate, as well as contaminant phases of the carbonates and oxides of other metallic elements. The contaminant phases may appear within the pores of the shell or form thinner layers on the inner surface of the shell [23].

The XRF imaging experiment was carried out at the P06 Hard X-ray Microprobe, PETRAIII, DESY, Germany, utilizing 17 keV monochromatic primary x-rays. This photon energy is much higher the K-edges of calcium (4.038 keV), iron (7.112 keV), and strontium (16.105 keV). The detector used was still a pnCCD. The optical component was a 6:1 magnifying conical polycapillary optic. The achieved spatial resolution could not exceed 8 μm at best. For the purpose of 3D imaging, forty-five XRF projections were taken at rotational angles between 0 and 180 degrees. The measurement time for each projection was 10 min. After collecting all the projections, 3D element maps were reconstructed using a conventional filtered back-projection algorithm. As shown in figure 2.6(f), which is the element map at the cross section indicated by the dashed line in figure 2.6(c), the distribution of calcium outlines the basic structure of the foraminiferal shell. The hollow

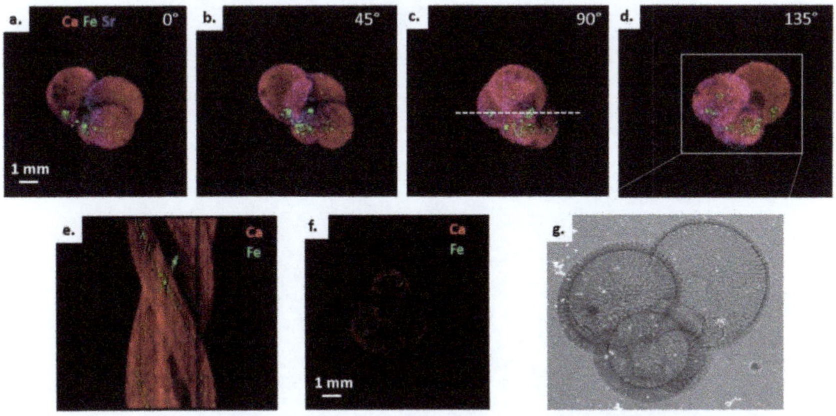

Figure 2.6. (a)–(d) Two-dimensional projections of calcium, iron, and strontium in a single foraminifer at rotational angles of 0, 45, 90, and 135 degrees, respectively. (e) A single sinogram of calcium and iron at the height indicated by the dashed line in (c). (f) A reconstructed virtual cross section through the foraminifer. (g) A photomicrograph of the foraminifer. Adapted from figure 8 in reference. Reproduced from [22] with permission from the Royal Society of Chemistry.

structure, which is not visible in a single 2D XRF projection, becomes evident. In addition, iron is enriched in certain spots on the shell.

This x-ray imaging technique can also be applied to other biological samples, such as plants. Figures 2.7–2.9 show the results of imaging bamboo leaves, cherry petals, and lilac petals, respectively [24]. The experiments were performed at the undulator beamline, BL-NW2A1, PF-AR (Photon Factory's Advanced Ring operated at 6.5 GeV), KEK, Japan. The detector used was a CCD camera (TC281 CCD, from Texas Instruments, 1000×1000 pixels, pixel size 8 μm, cooled to −30 °C). Prior to imaging, ordinary x-ray fluorescence spectra were measured in air, and it was found that the major element in the above plants is potassium (K). Imaging was then performed by simply integrating all the x-ray energy photons from the sample. The images were taken with an exposure time of 1 sec, repeated 30–50 times. So, the above x-ray images are almost maps of K. It was also found that the above samples contain some trace elements such as Mn, Fe, Cu, Zn, and Rb. Using single-photon counting with charge-sharing correction, it is possible to obtain the image specific to each element.

From the above examples, we can see the utility of XRF imaging in the field of biology. In XRF imaging, unlike electron microscopy, it is obvious that these biological samples do not require complex pretreatment, fixation, or a conductivity change. Meanwhile, projection-type XRF imaging can be used with samples that have uneven surfaces or simple fixation, while scanning-type XRF imaging requires the sample surface to be relatively flat and the sample to be securely fixed to prevent loosening or detachment during *x*–*y* scanning.

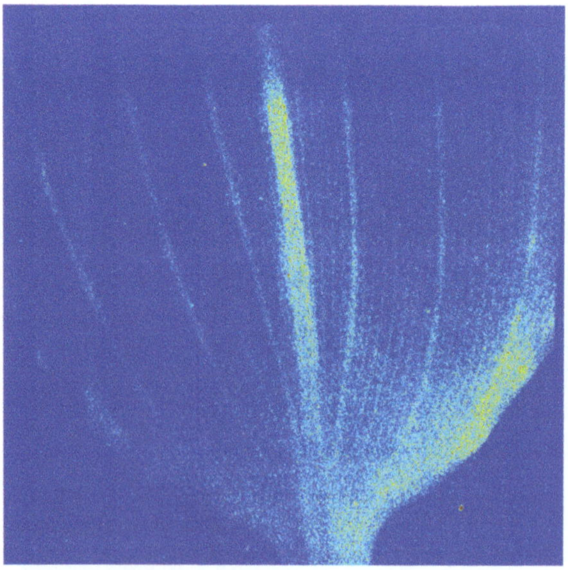

Figure 2.7. An XRF image of a bamboo leaf [24]. The major element is K. Excitation energy 9.03 keV. Exposure time 1 s. Repeated 30 times.

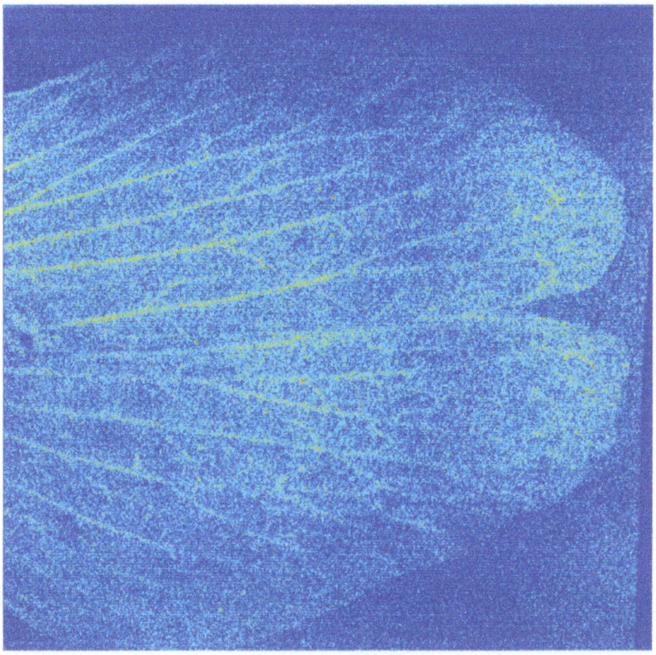

Figure 2.8. An XRF image of a cherry petal [24]. The major element is K. Excitation energy 7.2 keV. Exposure time 1 s. Repeated 50 times.

Figure 2.9. An XRF image of a lilac petal [24]. The major element is K. Excitation energy 7.2 keV. Exposure time 1 s. Repeated 30 times.

2.3 Mineralogy

Let us start this section by addressing a common problem in daily life: how to quickly distinguish diamonds, glass, and synthetic cubic zirconia (ZrO_2, a common substitute for diamonds in jewelry). Certainly, there are various methods available. In the meantime, for those familiar with XRF analysis, this problem is quite simple. From the standpoint of chemical composition, diamonds are made up solely of carbon; their carbon Kα fluorescence x-rays have a low energy of 0.277 keV. X-ray photons with such a low energy are strongly absorbed in air and therefore cannot be detected in the atmosphere. On the other hand, cubic zirconia consists of oxygen and zirconium. Due to the low energy of oxygen Kα emission (0.525 keV), oxygen is also undetectable in the atmosphere. However, zirconium can emit high-energy Kα (15.774 keV) and Kβ (17.666 keV) fluorescence x-rays, making it easy to detect and identify. As for glass, its primary component is silicon dioxide. Silicon Kα fluorescence x-rays have an energy of 1.74 keV and can be detected if the XRF detector is placed very close to the glass sample to minimize air absorption. During the production of glass, potassium oxide and calcium oxide are often added as fluxing agents. The introduced elements, potassium (Kα energy 3.313 keV) and calcium (Kα energy 3.691 keV), can be readily identified in XRF analysis. In addition, manufacturers sometimes add a large amount of lead to glass to increase its refractive index, giving it a sparkling appearance similar to that of diamonds. The introduced lead (Lα_1 energy 10.55 keV) is also easily detectable.

Certainly, distinguishing diamonds, glass, and cubic zirconia is a relatively simple task. They can also be distinguished using other routine identification methods. However, this simple example highlights the fact that gems or crystals with a similar appearance may possess entirely distinct chemical compositions, which can be readily identified through XRF analysis. This principle can also be extended to rocks and minerals. Obviously, natural minerals are mostly inhomogeneous, making it challenging to obtain characteristic information solely by analyzing the averaged elemental composition across a wide area. In such situations, the utilization of XRF imaging becomes imperative.

To illustrate the role of XRF imaging in minerology, we provide a simple example of turquoise, an opaque mineral exhibiting green to blue coloration. Its primary composition is $CuAl_6(PO_4)_4(OH)_8{\cdot}4H_2O$. Certain rare variations of turquoise with distinct hues can hold high value in the market. Typically, turquoise is found embedded within its host rock, retaining speckled or veined textures that often consist of pyrite or limonite. Consequently, when employing XRF imaging to analyze turquoise, a copper background and an iron pattern should be present. Geologically, turquoise deposits form through various mechanisms, leading to diverse patterns and the presence of other trace elements such as zinc or calcium [25]. The elemental map revealed by XRF imaging has the potential to establish a standard for turquoise evaluation.

On the other hand, in certain deposits, turquoise can coexist with variscite, which primarily comprises $AlPO_4{\cdot}2H_2O$. Due to their similar appearances, distinguishing them with the naked eye is challenging [26]. However, XRF imaging easily solves this issue, since variscite lacks a significant amount of copper, and its color mainly stems from the incorporation of vanadium and chromium elements.

Furthermore, XRF imaging and elemental mapping can serve as a valuable tool for identifying mineral zonations and mineral phases that offer insights into hydrothermal activity, fluid chemistry, and fluid–rock reactions. This information plays a crucial rule in comprehending ore formation processes and hydrothermal alternation. Moreover, variations in trace elements can serve as records of changes in physicochemical conditions during mineral growth. These analyses have been conducted to investigate the mechanisms behind gold emplacement [27] and the deposition of corallite concretion [28].

As a first example, figure 2.10 depicts the XRF imaging of a chatoyant gemstone known as tigereye [29]. Tigereye is a widely recognized and beloved gemstone. Its quartz matrix (SiO_2) contributes to its silky lustrous appearance. The coloration of tigereye arises from iron compounds, including yellowish-brown limonite [FeO (OH)$\cdot H_2O$] and red hematite (Fe_2O_3).

The XRF imaging experiment was carried out using a laboratory x-ray tube with a copper anode. To locate the viewing area, a white marker made of titanium dioxide was pasted onto the surface of the sample. In this experiment, the tube was operated at 40 kV and 35 mA. The primary x-rays were monochromatized to approximately 8.05 keV, which is the energy of the characteristic copper Kα radiation, using graphite (002) reflections. This energy is higher than the K-edges of titanium (4.965 keV), manganese (6.540 keV), and iron (7.112 keV). The detector was a conventional CCD camera (CCD47–10, from Teledyne e2v, 1024 $\times$ 1024 pixels, pixel size 13 μm, sensor thickness $\sim$ 10 μm, cooled to -30 °C). In this experiment, the CCD camera was operated in single-photon-counting mode and achieved an energy resolution of 150 eV at 5.9 keV. The x-ray optical component was a pinhole 38 μm in diameter manufactured on a 30 μm thick tungsten foil. The imaging magnification was 4 times, and the achievable spatial resolution was 50 μm. The total measurement time was 2 h. Elemental maps of titanium, manganese, and iron were generated utilizing the XRF signals of titanium Kα (4.510 keV) and Kβ (4.931 keV), manganese Kα (5.898 keV), and iron Kα (6.403 keV) and Kβ (7.057 keV), respectively.

Upon comparing the elemental maps with the photomicrograph, it becomes evident that the dark brown fibers within the sample align with iron, suggesting that it is likely present in the form of limonite. In addition, two black spots indicate the presence of manganese. Actually, it is quite common to observe the inclusion of manganese in limonite, either as discrete manganese oxide nanocrystals or as a substitute for iron within a goethite (FeO(OH)) lattice [30].

The second example involves the XRF imaging of a gabbro [31], as shown in figure 2.11. The XRF imaging experiment was carried out at BL-14B, Photon Factory, KEK, Japan, utilizing monochromatic primary x-rays with an energy of 9.53 keV, which is higher than the K-edges of calcium (4.038 keV), titanium (4.965 keV), manganese (6.540 keV), and iron (7.112 keV). The detector was the same CCD camera as that utilized for figure 2.10. It was operated in single-photon-counting mode. The optical component was the same collimator plate as that utilized for

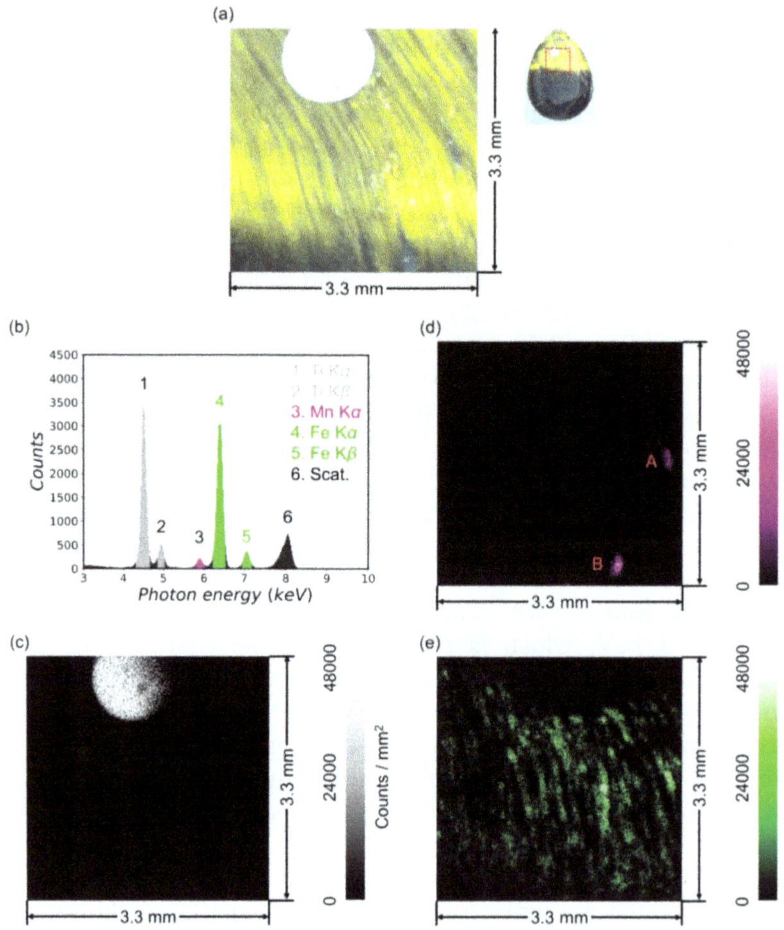

Figure 2.10. XRF imaging of the gemstone tigereye. (a) Optical photo of the sample. A white mark of titanium dioxide was pasted on its surface. (b) XRF spectra of the 3.3 mm × 3.3 mm viewing area. (c)–(e) Maps of titanium (Ti Kα and Kβ), manganese (Mn Kα), and iron (Fe Kα and Kβ), respectively. Modified from figure 5 in reference. Reprinted from [29], with the permission of AIP Publishing.

figure 2.2. The total measurement time was 8.3 h. Elemental maps of calcium and iron were generated utilizing the XRF signals of calcium Kα (4.510 keV) and Kβ (4.931 keV) and iron Kα (6.403 keV) and Kβ (7.057 keV), respectively. For this experiment, the instrument layout was specially designed to position the detector in the polarization direction of the primary x-rays. This arrangement aimed to minimize the scattering background. Detailed discussions of this instrument layout can be found in section 3.2.

This gabbroic sample was collected from the summit of Mount Tsukuba, a solitary mountain situated on the northeastern edge of the Kanto Plain in Japan. The gabbroic rocks in this mountain were formed by erosion approximately 7600 million years ago. In a previous study [32] of a similar gabbroic rock collected from

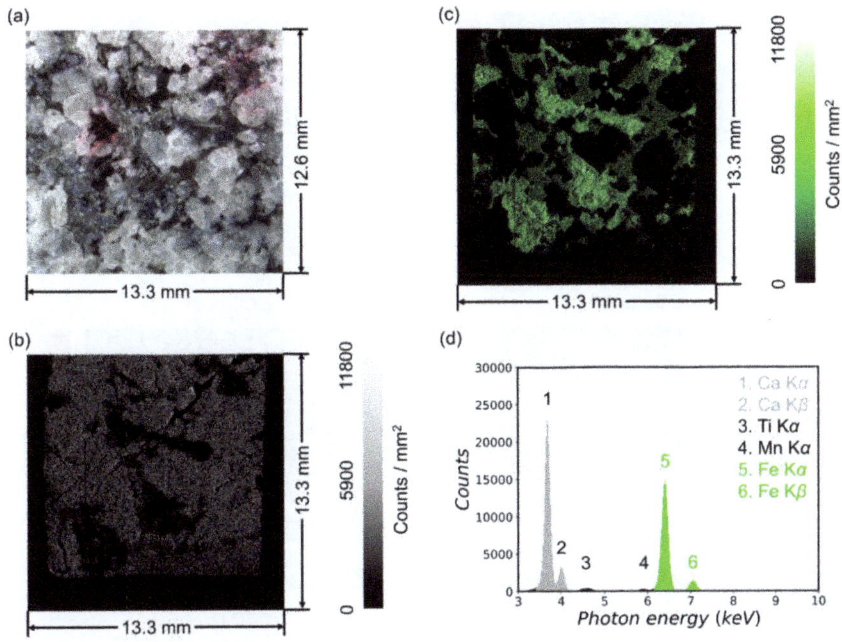

Figure 2.11. XRF imaging of a gabbro collected from Mount Tsukuba, Japan. (a) An optical photo of the sample. (b), (c) Maps of calcium (Ca Kα and Kβ) and iron (Fe Kα and Kβ), respectively. (d) XRF spectra of the 13.3 mm × 13.3 mm viewing area. Modified from figure 2 in reference. Reprinted from [31], with the permission of AIP Publishing.

the same location, it was found that the white matrix primarily consists of feldspar, specifically anorthite ($CaAl_2Si_2O_8$), which accounts for over 10% of the sample's weight in calcium. The black textures, on the other hand, contain a combination of amphiboles and olivine. The olivine is a solid solution consisting of 50% forsterite (Mg_2SiO_4) and 50% fayalite (Fe_2SiO_4). The weight percentage of iron in these black textures exceeds 10%. This XRF imaging experiment successfully revealed the distributions of calcium and iron.

The third example, as shown in figure 2.12, involves the XRF imaging of lapis lazuli [17]. The sample surface was non-flat. The XRF imaging experimental conditions were the same as those of figure 2.3, except that the laboratory x-ray tube was operated at 40 kV and 20 mA, and the total measurement time was 2700 s. It is worth noting that compared to the high-energy barium K-line fluorescence, measuring and analyzing the low-energy barium L-line fluorescence is much easier. The L_1-edge, L_2-edge, and L_3-edge of barium are 5.989, 5.623, and 5.247 keV, respectively, which are much lower than the characteristic radiation of molybdenum Kα (17.478 keV) and Kβ (19.607 keV) in the primary x-rays. The energy of the strongest barium $L\alpha_1$ is 4.466 keV.

Lapis lazuli is a type of a semi-precious stone that has been highly valued throughout history for its rich, deep blue color. Its origins can be traced back to as early as 7570 BC at the oldest site of the Indus Valley civilization, Bhirrana. Lapis

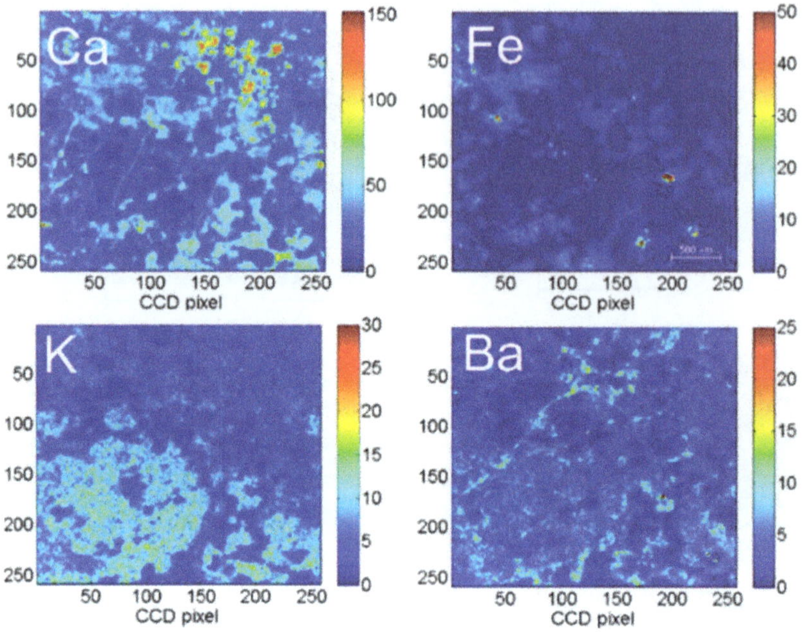

Figure 2.12. Elemental maps of lapis lazuli from Siberia. Reprinted from [33], with permission from Elsevier.

lazuli can be crafted into a variety of items, including jewelry, carvings, mosaics, small statues, and vases. In medieval Europe, it was ground to make the natural ultramarine pigment, which became the most expensive pigment of that time.

Although lapis lazuli ($(Na,Ca)_8(AlSiO_4)_6(S,SO_4,Cl)_{1-2}$) can sometimes be mistaken for azurite ($Cu_3(CO_3)_2(OH)_2$) or sodalite ($Na_8(Al_6Si_6O_{24})Cl_2$) based on its appearance, XRF analysis can easily distinguish them. The vibrant color of lapis lazuli is attributed to the trisulfur radical anion within its crystal structure, while azurite's blue color is due to copper hydroxycarbonate and requires a high copper content. On the other hand, the presence of pyrite inclusions can be utilized to distinguish lapis lazuli from sodalite. In the elemental maps of lapis lazuli, we should be able to observe iron-rich islands. In the case of the lapis lazuli specimen presented in figure 2.12, its Siberian origin [33] can be recognized by the combination of calcium-rich and potassium-rich phases, along with traces of barium and iron inclusions (in the form of pyrite, FeS_2).

From the examples provided above, we can see the practical implementation of XRF imaging in mineral analysis. It should be noted that while EPMA can also be utilized to map the elements in ores, its analysis is always limited to a skin depth of 1–2 microns on the sample, which may not provide sufficient information to characterize the texture. In addition, unlike scanning-type XRF imaging, projection-type XRF imaging does not necessitate a complex *x*–*y* positional scanner. It can employ compact, portable instruments, enabling on-site mineral analysis during field surveys. This will significantly enhance the efficiency of mineral survey work.

2.4 Industry

XRF imaging has a wide range of industrial applications. In this section, we will begin by exploring its usage in inspecting the quality of circuit boards. The thin wires present in circuit boards are typically hidden within multiple layers, making them inaccessible for examination under an optical microscope. X-ray radiography, also known as x-ray transmission imaging, can penetrate electrical components and reveal their internal structures non-destructively with high imaging efficiency. However, the contrast in x-ray transmission images primarily originates from different mass thickness densities, making it unable to accurately depict the shapes of specific components made from particular materials. XRF imaging can also be applied to inspect circuit boards. It has the ability to distinguish between different components made of various chemical elements. Consequently, it can clearly identify copper wires, soldering materials, and other high-density parts within the electrical components, which are only indirectly seen in x-ray radiography.

Figure 2.13 shows the XRF imaging of a circuit board [34]. The XRF imaging experiment was carried out at BL-16A1, Photon Factory, KEK, Japan, utilizing monochromatic primary x-rays with an energy of 10 keV, which is higher than the copper K-edge (8.979 keV). The detector was a conventional CCD camera (TC281 CCD, from Texas Instruments, 1000×1000 pixels, pixel size 8 μm, cooled to −30 °C) that originally lacked the ability to resolve x-ray energies. The x-ray optical component was the same collimator plate as that utilized for figure 2.2. The exposure time for the XRF image in figure 2.13(b) was 200 milliseconds. As the utilized detector could not resolve x-ray energies, the intensity in the XRF image may come from the XRF signals of multiple elements that can be excited by the 10 keV primary x-rays. However, based on our prior knowledge of the circuit board, we can affirm that the pattern in the XRF image corresponds to copper wires. It is

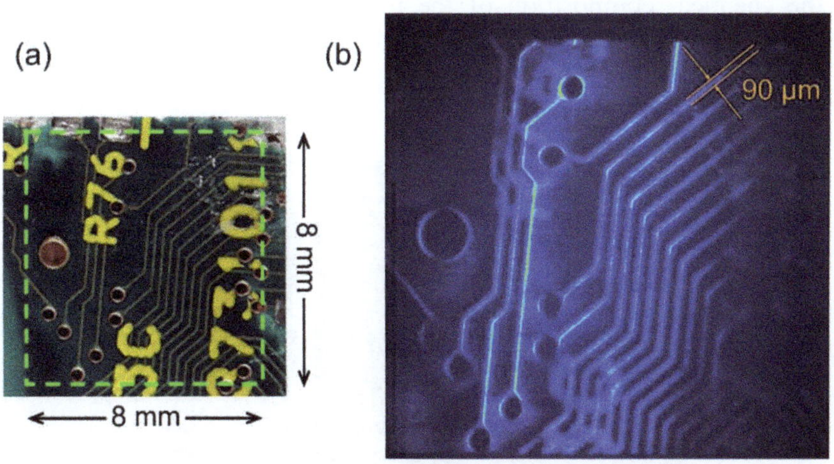

Figure 2.13. The XRF imaging of printed wires on an exposed circuit board. (a) Optical photo. The dashed rectangle encloses the viewing area. (b) The XRF image, reprinted from [34], with permission of AIP Publishing.

worth noting that, in this experiment, a high-quality XRF image was obtained in just 200 milliseconds, allowing the fine and dense wires in the circuit board to be clearly visible. This rapid measurement makes XRF imaging highly suitable for swift inspections of products on factory assembly lines.

To accurately identify different metal components in circuit boards, such as copper wires, silver wires, solders composed of lead and tin, and mechanical components such as screws made of stainless steel, we can utilize detectors that can resolve x-ray energies. Figure 2.14 shows the XRF imaging of copper in an enclosed microSD card [35]. The circuit board was concealed beneath a plastic cover. The XRF imaging experiment was carried out at BL-14B, Photon Factory, KEK, Japan, utilizing monochromatic primary x-rays with an energy of 13.5 keV, which is higher than the copper K-edge (8.979 keV). The detector was a visible-light scientific CMOS camera (PCO.edge 5.5 sCMOS made by PCO AG, 2560 × 2160 pixels, pixel size 6.5 μm, cooled to 5 °C). In this experiment, the scientific CMOS camera was operated in single-photon-counting mode and achieved an energy resolution of 220 eV at 5.9 keV. The x-ray optical component was a pinhole 50 μm in diameter manufactured on a 30 μm thick tungsten foil. The imaging magnification was approximately 4.4 times. The total measurement time was 10 min. The map of copper was generated using copper Kα (8.047 keV) fluorescence x-rays. The revealed structure of copper wires is quite typical, similar to that in most microSD cards.

The second example is more interesting; it concerns the use of XRF imaging on the production lines of food factories to detect foreign hazardous fragments in food. These fragments may accidentally detach from the equipment on the production line or originate from raw materials that have not undergone thorough examination. X-ray radiography can be a highly effective and noninvasive inspection method. On the other hand, XRF imaging offers additional information about the chemical composition of the foreign fragments. This information is crucial for determining the type of fragments and inferring their possible sources.

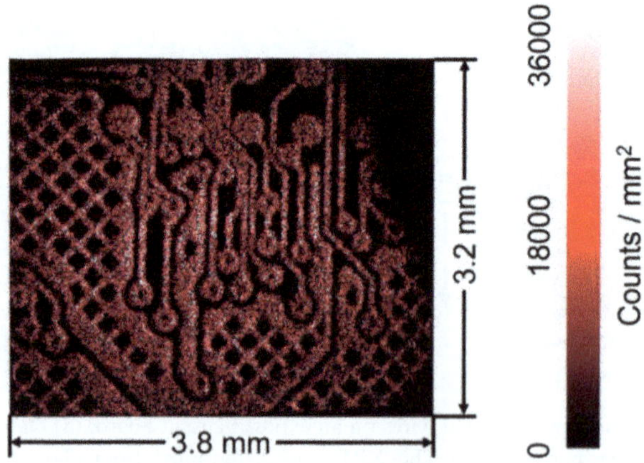

Figure 2.14. An XRF image of copper in an encapsulated microSD card [35].

For example, the majority of fragments could be ceramic (glass) or steel. While these fragments may display a similar appearance of strong absorption in x-ray transmission images, their XRF spectra are completely distinct when observed using color XRF imaging.

Figure 2.15 shows the XRF imaging of metal and glass fragments buried in a cheesecake [35]. The XRF imaging experiment was carried out using monochromatized primary x-rays from a laboratory x-ray tube with a copper anode. In this experiment, the tube was operated at 40 kV and 300 mA. The primary x-rays were monochromatized to approximately 8.05 keV, which is the characteristic energy of copper Kα radiation. This energy is higher than the K-edges of potassium (3.607 keV), calcium (4.038 keV), chromium (5.989 keV), and iron (7.112 keV). The detector was the same visible-light scientific CMOS camera as that for figure 2.14 and was also operated in single-photon-counting mode. The x-ray optical component was a pinhole 100 μm in diameter manufactured on a 30 μm thick tungsten foil. The imaging magnification was approximately 1.73 times, and the achievable spatial resolution was 160 μm. The total measurement time was set to 1 h to accumulate a strong signal. However, it is also possible to reduce the time to 10 min if rapid inspection is preferred.

In the XRF spectrum of the entire imaging area shown in figure 2.15(e), strong peaks corresponding to calcium, chromium, and iron are clearly visible and are sufficient to trigger an alarm indicating the presence of foreign fragments. The maps of calcium, chromium, and iron were generated using the XRF signals of calcium Kα (3.691 keV), chromium Kα (5.414 keV), and iron Kα (6.403 keV), respectively, as shown in figures 2.15(b)–(d). These maps indicate the location and outline of the fragments. Furthermore, analyzing the XRF spectra in the regions of the fragments helps to determine their composition. In reality, the fragment at region A is glass, and therefore its XRF spectrum predominantly corresponds to calcium. The fragment at region B is stainless steel SUS309S (standard reference material JSS653-14), which contains 12%–15% nickel and 22%–24% chromium. In the XRF spectrum of region B, nickel peaks are not visible because the photon energy of the primary x-rays (8.047 keV) is lower than the nickel K-edge (8.333 keV). However, there are prominent peaks of chromium and iron, suggesting that the fragment could be stainless steel. In comparison, the fragment at region C is low-alloy steel (standard reference material JSS151-16), which contains ~1.7% manganese and ~2.9% nickel. Therefore, its XRF spectrum displays prominent iron peaks and a small manganese peak.

A similar example of a chocolate is shown in figure 2.16 [35]. The experimental conditions were exactly the same as those of figure 2.15. The fragment was still stainless steel SUS309S (standard reference material JSS653-14). Elemental maps of potassium, calcium, chromium, and iron were generated using the XRF signals of potassium Kα (3.313 keV), calcium Kα (3.691 keV), chromium Kα (5.414 keV), and iron Kα (6.403 keV), respectively, as shown in figures 2.16(c)–(f). It is easy to notice the homogeneous distribution of potassium and calcium, indicating that the potassium and calcium signals originate from the chocolate itself rather than from

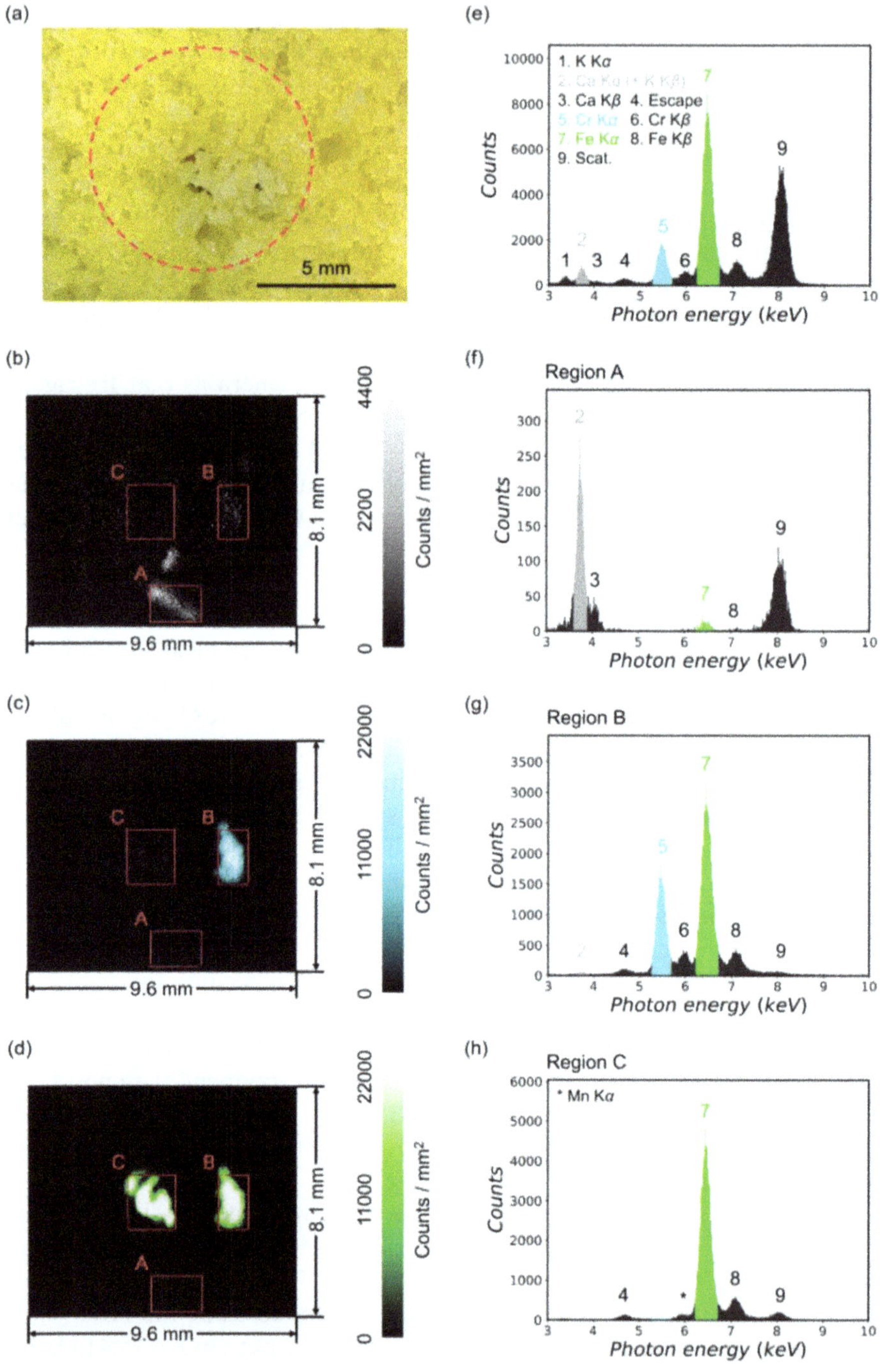

Figure 2.15. XRF imaging of metal and glass fragments in a cheesecake [35]. (a) An optical photo of the cheesecake. Fragments are invisible from the surface. The dashed circle encloses the area of XRF imaging. (b)–(d) Images corresponding to calcium (Ca Kα), chromium (Cr Kα), and iron (Fe Kα), respectively. (e)–(h) The XRF spectra of the entire imaging area for region A, region B, and region C, respectively.

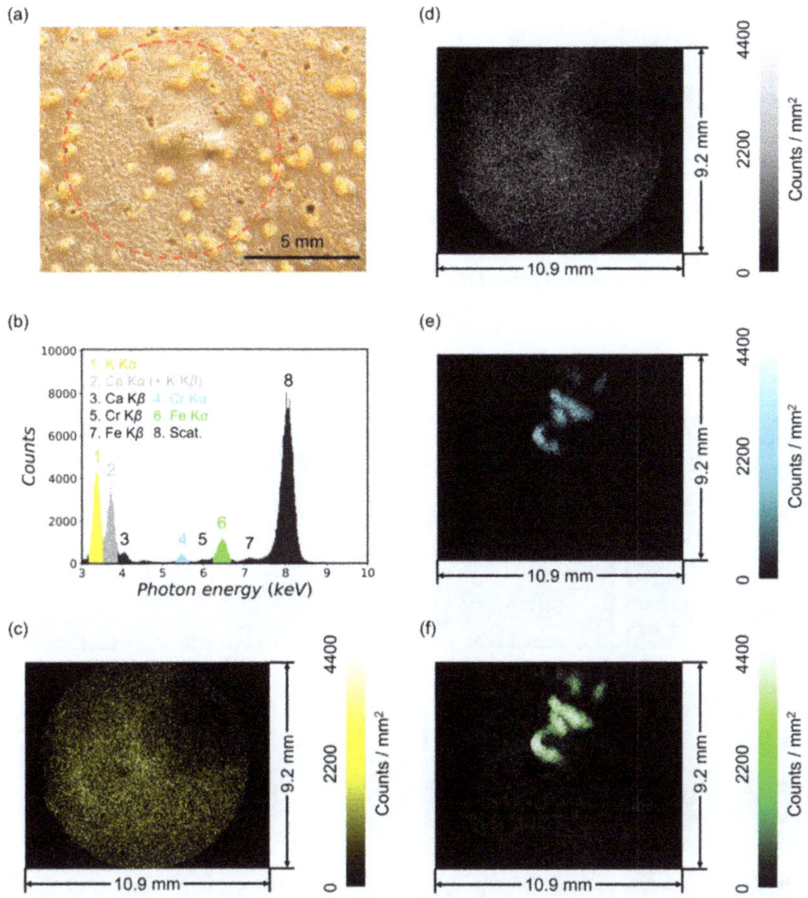

Figure 2.16. XRF imaging of a stainless steel fragment buried in a chocolate [35]. (a) An optical photo of the chocolate. The dashed circle encloses the viewing area of the XRF imaging. (b) The XRF spectrum of the entire viewing area. (c)–(f) Maps of potassium (K Kα), calcium (Ca Kα), chromium (Cr Kα), and iron (Fe Kα), respectively.

foreign fragments. This once again demonstrates that XRF imaging is more informative than ordinary XRF analysis.

The third example is the XRF imaging of a printed paper (see figure 2.17) [35]. The experimental conditions were the same as those of figure 2.11, except that the total measurement time was 1.7 h. The maps of iron were generated using the XRF signals of iron Kα (6.403 keV) and Kβ (7.057 keV). In the map of iron, the text is easily discernible, since the ink contains iron. In the past, people commonly utilized iron gall ink for writing. Although iron gall ink is no longer widely used today, modern inks still incorporate diverse metallic elements in their colorants and additives. This characteristic makes XRF imaging applicable to printed materials. In certain scenarios, XRF imaging can differentiate between various ink types used in printing or aid in the recovery of obscured or altered documents.

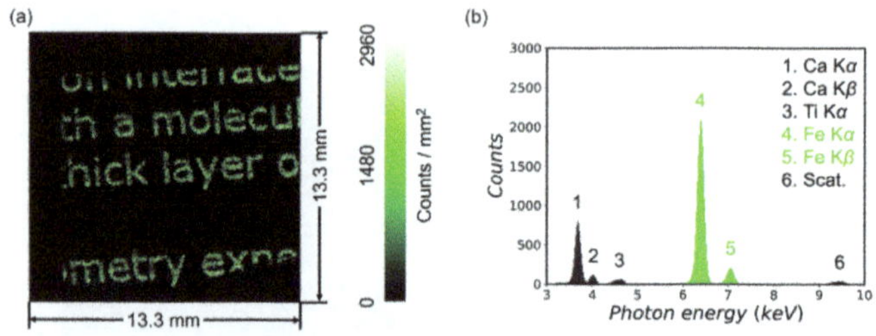

Figure 2.17. XRF imaging of a printed paper [35]. (a) A map of iron (Fe Kα and Kβ). (b) The XRF spectra of the sample.

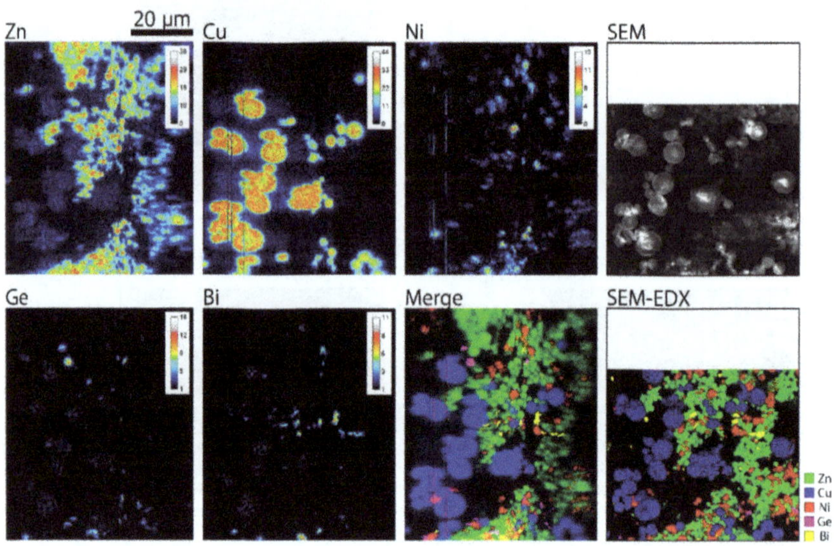

Figure 2.18. XRF images of micron-sized particles of nickel dioxide, copper, zinc, germanium, and bismuth. Scanning electron microscopy (SEM) and SEM with energy-dispersive x-ray spectroscopy (SEM–EDX) images are also shown. Adapted with permission from figure 6 in reference [36], copyright The optical Society.

For the fourth example, we show the XRF imaging of a mixture of various micron-sized particles in figure 2.18 [36]. This artificial sample was initially made to demonstrate the submicron resolution achievable with XRF imaging. However, this example also showcases the powerful capability of XRF imaging for analyzing disordered powdered samples, which holds potential for applications in the chemical industry or forensic investigations.

The XRF imaging experiment was carried out at BL29XUL, SPring-8, Japan, utilizing monochromatic primary x-rays with an energy of 20 keV, which is higher than the K-edges of nickel (8.333 keV), copper (8.979 keV), zinc (9.659 keV), and germanium (11.104 keV), as well as the L_1-edge of bismuth (16.385 keV). The detector was a conventional CCD camera (PI-LCX: 1300, from Princeton Instruments, 1340×1300 pixels, pixel size 20 μm, cooled to −100 °C) that originally lacked energy resolution. In this experiment, the CCD camera was operated in single-photon-counting mode and achieved an energy resolution of 180 eV at 8.15 keV. The x-ray optical component was a pair of advanced Kirkpatrick–Baez (AKB) mirrors. The imaging magnification was 25.7 times in the vertical direction and 83.5 times in the horizontal direction. The spatial resolution achieved was 500 nm, which is the best performance of projection-type XRF imaging so far. The total measurement time was 3.3 h.

The scenarios and demands of industry are always diverse. It is impossible for us to enumerate all the possible applications of projection-type XRF imaging in industry. In this chapter, we only present a few illustrative examples. However, it is commonly understood that in industrial applications, people often need to utilize the penetrative power of x-rays to examine the internal structures of samples. This poses a challenge for EPMA, which primarily focuses on surface analysis. In addition, in industrial production, there is often a need to quickly and timely inspect samples in coordination with the assembly line. The inspection equipment itself needs to be robust and adaptable to samples of various shapes, sizes, and conditions. This makes projection-type imaging, which utilizes a stationary instrument layout, preferable to scanning-type XRF imaging that requires a complex x–y positional sample scanner.

2.5 Cultural heritage

Over the last twenty years, XRF imaging has played a significant role in the study of cultural heritages and artworks, with a particular focus on paintings. This application is based on the chemical elements present in inorganic pigments. Table 2.2 lists the chemical compositions of some commonly used inorganic pigments, along with brief remarks on their historical succession or alternation. As the number of inorganic pigment species extends beyond those listed in the table, and their history can be quite complex, readers are encouraged to explore further details in books on art history.

It is widely understood that the chemical compositions of pigments of the same color can vary significantly across different regions or eras. This distinction provides an important clue for identifying paintings [37–40]. Moreover, the displayed colors of inorganic pigments can change over time due to various chemical reactions. One well-known example is lead white, which can darken as a result of an oxidation reaction that coverts divalent lead to tetravalent lead in lead dioxide, or due to a reaction with small traces of hydrogen sulfide in the air, resulting in the formation of black lead sulfide. These reactions also account for the darkening of minimum (lead red) and certain other lead-containing pigments [41].

In reality, the degradation of color appearance due to exposure to light, moisture, acid, alkali, and organic pigments is a highly intricate process. However, the deposition of metallic elements remains unchanged. The element maps obtained through XRF imaging can help to uncover the original color appearance. In addition, pigments on the surface of cultural heritage may erode or be scraped away, rendering the original colors invisible. However, a small amount of pigment

Table 2.2. The chemical compositions of some common inorganic pigments.

Color	Pigment name	Chemical formula	Remarks
Black	Carbon black	C	Used since ancient times
	Mars black	Fe_3O_4	Developed in the early 20th century
	Manganese dioxide	MnO_2	Used in Paleolithic Lascaux cave paintings; not popular in oil paintings
White	Lead white	$(PbCO_3)_2 \bullet Pb(OH)_2$	The principal white pigment from ancient times until the 20th century; toxic
	Barium sulfate	$BaSO_4$	Introduced as a pigment in the late 18th century
	Lithopone	ZnS (30%) and $BaSO_4$ (70%)	First produced in the 1870s
	Zinc white	ZnO_2	Used in watercolor from the 1830s; became popular in oil paintings from the last quarter of the 19th century
	Titanium white	TiO_2	Introduced as a pigment in the early 20th century; anatase, rutile, or brookite
	Antimony white	Sb_2O_3	Introduced as a pigment in 1920; toxic
Red	Red ocher	Fe_2O_3	Used in Paleolithic Lascaux cave paintings; widely used since ancient times
	Realgar	As_2S_2	An ancient pigment; used in European paintings during the Renaissance era; use ended in the 18th century; toxic
	Minimum	Pb_3O_4	Named as 'cinnabar of lead' in the Han dynasty of China; commonly used in the Middle Ages in Europe; toxic
	Vermilion (cinnabar)	HgS	Widely used in ancient China; the primary red pigment in Europe from the Renaissance until the 20th century, toxic
	Cadmium red	Cd(S,Se)	Introduced as a pigment in 1907; toxic
Purple	Han purple	$BaCuSi_2O_6$	Found in paintings and ceramic glazes in the Han dynasty of China
	Purple of Cassius	$Au_x \cdot SnO_2$	First made in the second half of the 17th century; used until the mid-19th century; expensive
	Cobalt violet	$Co_3(PO_4)_2$	First made in 1859; the first real violet pigment
	Manganese violet	$NH_4MnP_2O_7$	First made in 1866; not widely used
Yellow	Orpiment	As_2S_3	Used in many early civilizations; in Europe, widely used from the 16th century until the 19th century; toxic

	Lead-tin yellow	Pb_2SnO_4 / $Pb(Sn,Si)O_3$	In use from the beginning of the 14th century; widely used in the Renaissance; use ended at the end of the 17th century; toxic
	Naples yellow	$Pb_2Sb_2O_7$	Frequently used in paintings during the period 1750–1850; toxic
	Chrome yellow	$PbCrO_4$	Introduced as a pigment in the early 1800s; toxic
	Cobalt yellow	$K_3Co(NO_2)_6$	Introduced as a pigment in 1852; remained popular until the late 19th century
	Cadmium yellow	CdS	First synthesized in 1817; became popular in the early 20th century; toxic
	Bismuth vanadate	$BiVO_4$	First synthesized in 1924 and became popular after the 1980s
	Titanium yellow	$NiO{\cdot}Sb_2O_3{\cdot}20TiO_2$	Introduced as a pigment in 1954
	Lemon yellow	$SrCrO_4$ / $ZnCrO_4$ / $BaCrO_4$	Toxic
	Mosaic gold	SnS_2	Inexpensive gold-like pigment; 13th–19th century
Green	Green earth	$K[(Al,Fe^{3+}),(Fe^{2+},Mg)](AlSi_3,Si_4)O_{10}(OH)_2$	Ground from green colored siliceous, iron-rich clays; used since ancient times
	Malachite	$Cu_2CO_3(OH)_2$	Similar component to azurite (used in blue colorants)
	Rinman's green	$Zn_{1-x}Co_xO$	First synthesized in 1780
	Viridian	$Cr_2O_3{\cdot}xH_2O$	Introduced as a pigment in 1838
	Scheele's green	$CuHAsO_3$	Synthesized in 1775; not widely used as a paint pigment
Blue	Han blue	$BaCuSi_4O_{10}$	Found in paintings and ceramic glazes in the Han dynasty of China
	Egyptian blue	$CaCuSi_4O_{10}$	Used in ancient Egypt
	Ultramarine	$Na_{8-10}Al_6Si_6O_{24}S_{2-4}$	Ground from lapis lazuli; very expensive; synthesized in 1824
	Azurite	$Cu_3(CO_3)_2(OH)_2$	Widely used in European Renaissance paintings
	Prussian blue	$Fe_4[Fe(CN)_6]_3$	First synthesized in 1706; soon became popular in painting
	Cobalt blue	$CoAl_2O_4$	Introduced as a pigment in the early 19th century; used in ancient Chinese ceramics; the blue color comes from impure cobalt oxides
	Cerulean blue	$CoO{\cdot}nSnO_2$	First synthesized in 1805
	Manganese blue	$BaMnO_4{\cdot}BaSO_4$	Discovered in 1907; not used nowadays

may still remain on the matrix's surface. XRF imaging can sensitively detect the distribution of these remaining pigments, thereby revealing vanished historical information.

In addition to paintings, ceramics also showcase a wide variety of colors. The color mechanism of ceramics is even more intricate due to the complex chemical reactions that take place when the raw materials are sintered at high temperatures. The same element can exhibit a diverse range of colors depending on factors such as the sintering temperature, the use of an oxidizing or reducing atmosphere, and the element's interaction with other elements [42]. However, by comparing the element map obtained through XRF imaging with the visual appearance of colors, we can enhance our understanding of the ceramic sintering process and the raw materials employed.

One of the most famous stories in the history of art involves the use of XRF imaging to uncover the somber face of a provincial Dutch woman concealed beneath a vibrant, Parisian-style floral painting by Vincent van Gogh [43]. This discovery reaffirmed van Gogh's uncomfortable financial situation and his necessity to reuse canvas. It allowed for a comparison between this hidden face and van Gogh's other known works, providing insights into the evolution of his artistic style.

In the study of cultural heritage, XRF imaging is often combined with other modern spectroscopic techniques to obtain a comprehensive understanding of chemical substances. These techniques include infrared reflectography, Raman microspectroscopy, Fourier transform infrared microspectroscopy, secondary electron microscopy, microregion x-ray diffraction, and microregion x-ray absorption spectroscopy. Numerous scientific reports are available on these subjects, as listed in review articles [44, 45]. In the following, we provide specific instances of the application of projection-type XRF imaging.

Let us start with an example of an oil painting. Figure 2.19 shows a portion of Gustave Caillebotte's work, 'Portraits à la campagne' (portraits in the countryside) and the corresponding elemental maps [46]. Gustave Caillebotte (1848–1894) was a French Impressionist painter. This painting was created in 1876. It shows the painter's mother, aunt, cousin, and a family member. The 35 mm × 35 mm viewing area of XRF imaging includes his mother's face as well as the background of flowers.

The XRF imaging experiment was carried out utilizing non-monochromatized primary x-rays from two laboratory x-ray tubes with palladium anodes. In this experiment, the tubes were operated at 30 kV and 100 μA. The primary x-ray energies included the characteristic radiation of palladium Kα (21.175 keV) and Kβ (23.816 keV), both of which have energies higher than the K-edges of calcium (4.038 keV), chromium (5.989 keV), iron (7.112 keV), and cobalt (7.709 keV), as well as the L_1-edges of mercury (14.846 keV) and lead (15.860 keV). The detector was a conventional CCD camera (iKon-M934, made by Andor, 1024×1024 pixels, pixel size 13 μm, sensor thickness 40 μm, cooled to −80 °C) that originally lacked energy resolution. In this experiment, the CCD camera was operated in single-photon-counting mode and achieved an energy resolution of 180–200 eV at 5.9 keV. The x-ray optical component was a square-pore microchannel plate 1.2 mm thick with

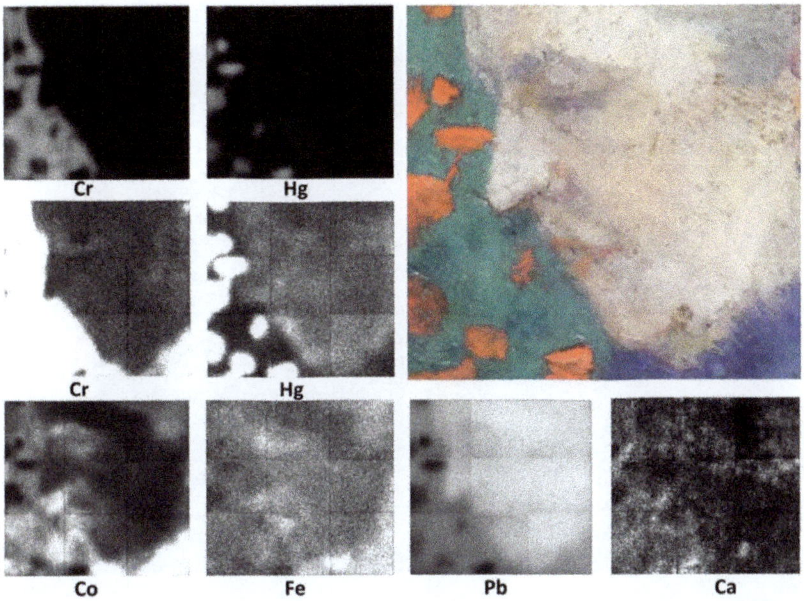

Figure 2.19. Elemental maps of a portion in Gustave Caillebotte's work, 'Portraits à la campagne' (portraits in the countryside). The viewing area used for XRF imaging was 35 mm × 35 mm. Adapted from figure 6 in reference [46], copyright 2018 John Wiley & Sons Ltd.

20 μm × 20 μm microchannels; the plate was capable of projecting 1:1 magnification images at a focal distance of 50 mm either from the sample or from the detector. The spatial resolution achieved was 200 μm. The total measurement time was 7.5 h (figure 2.19).

As depicted in figure 2.19, chromium and mercury are presented with two levels of contrast to demonstrate their use in thick layers or in thin brushstrokes. By referring to the relationship between colors, pigments, and chemical formulas in table 2.2, we can readily grasp the connection between the distribution of elements and the color appearance in the painting. Chromium corresponds to chrome green (viridian, $Cr_2O_3 \cdot xH_2O$), which was heavily used in the flowerbed, and small amounts of chrome yellow ($PbCrO_4$) used for the face. Mercury corresponds to red vermilion (HgS), which was heavily used in the flowers and lips and used in moderation to adjust the facial coloration. Cobalt corresponds to cobalt blue ($CoAl_2O_4$), which was applied to the clothing and facial contours. Iron corresponds to red ocher (Fe_2O_3) or brown earth pigment (the latter of which is a natural earth pigment not listed in table 2.2). Lead corresponds to lead white [$(PbCO_3)_2 \cdot Pb(OH)_2$]. Calcium corresponds to various compounds on the surface of the painting. It is worth noting that the creation date of this work in 1876 agrees with the primary period of usage for the corresponding pigments listed in table 2.2.

The second example is the XRF imaging of a painted pottery fragment [47], as shown in figure 2.20. The sample is from the Nazca culture in Peru, dating back to the 5th century AD. The Nazca culture is characterized by its polychrome pottery.

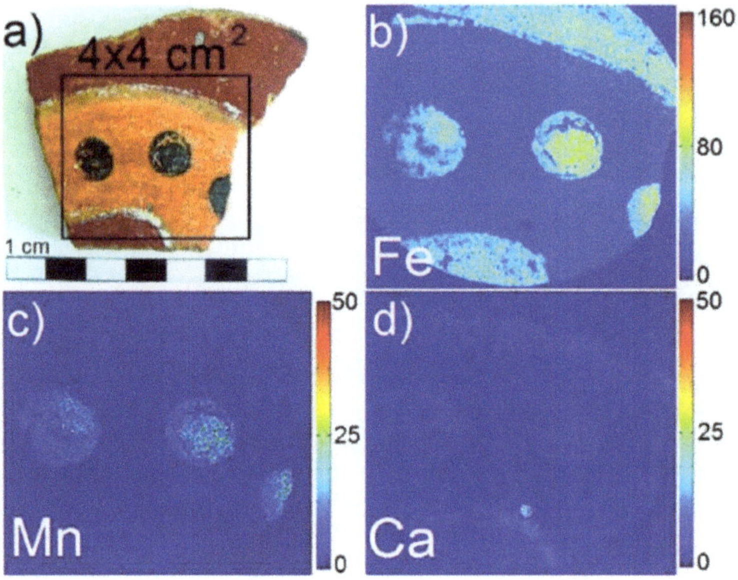

Figure 2.20. XRF imaging of a Nazca painted pottery fragment. (a) An optical photo of the non-flat sample. The rectangle encloses the area visible in the XRF imaging. (b)–(d) Elemental maps of iron, manganese, and calcium, respectively. Adapted with permission from figure 9 in reference [47], copyright (2014) American Chemical Society.

Mineral pigments were painted onto the slip before firing. The XRF imaging experiment was carried out utilizing non-monochromatized primary x-rays from a laboratory x-ray tube with a tungsten anode. The maximum power of the x-ray tube was 50 kV at 2 mA. This tube voltage is lower than the tungsten K-edge (69.524 keV), and therefore the characteristic radiations of tungsten Kα and Kβ could not be emitted. The primary x-rays consisted of bremsstrahlung radiation with continuous energy variation up to 50 keV. The detector was a conventional CCD camera (made by Andor, 1024×1024 pixels, pixel size 13 μm, sensor thickness 40 μm, cooled to −100 °C) that originally lacked energy resolution. In this experiment, the CCD camera was operated in single-photon-counting mode and achieved an energy resolution of 180 eV at 5.9 keV. The optical component was a pinhole 50 μm in diameter manufactured on a 75 μm thick tungsten foil. The imaging magnification was 0.35 times, and the spatial resolution achieved was 170 μm. The total measurement time was 5000 s. The 4 cm × 4 cm area used for XRF imaging is enclosed by the solid rectangle in figure 2.20(a).

Upon comparing the sample's color appearance and the elemental maps, it is evident that the black color corresponds to manganese and iron. This is primarily due to the presence of spinel jacobsite ($MnFe_2O_4$), which is formed by firing manganese- and iron-rich clay slips under oxidizing conditions [48]. The red color is mainly due to the presence of red iron (Fe_2O_3), which also forms under oxidizing conditions during firing. The white color corresponds to calcium.

The third example is the XRF imaging of a tin–lead glazed majolica fragment from Caltagirone, Italy, dating back to the 17th century AD [47], as shown in figure 2.21. The XRF imaging experimental conditions were exactly the same as those of figure 2.20. The distribution of lead in the glaze layer appears to be uniform. In areas where blue and yellow patterns are present, the XRF intensity of tin decreases, indicating the utilization of a two-layer glazing technique. This technique involves first creating an inner layer of opaque glaze containing tin oxide, followed by a second firing of a colorful outer layer of glaze containing pigments. The blue color mainly corresponds to cobalt oxide, which is a common source of blue color in pottery and ceramics. The yellow color mainly corresponds to a mixture of Naples yellow ($Pb_2Sb_2O_7$) and ferric oxide (Fe_2O_3).

The fourth example is the XRF imaging of a 2800-year-old Phoenician ivory object [49]. The sample is a carved ivory object from Syria, dating back to the 8th century BC. It was attributed to the Arslan Tash site, which is famous for many ivory objects of high artistic quality. It is believed that these ivory objects were originally extensively colored with a combination of inorganic pigments and organic binders. However, the pigments have generally not been preserved due to the degradation of organic binders over time. Fortunately, with XRF imaging, it is now possible to detect the remaining trace elements, which are indicative of the corresponding pigments and the original colors. Figure 2.22(a) displays an optical image of a portion of a wave frieze. Elemental maps are displayed in

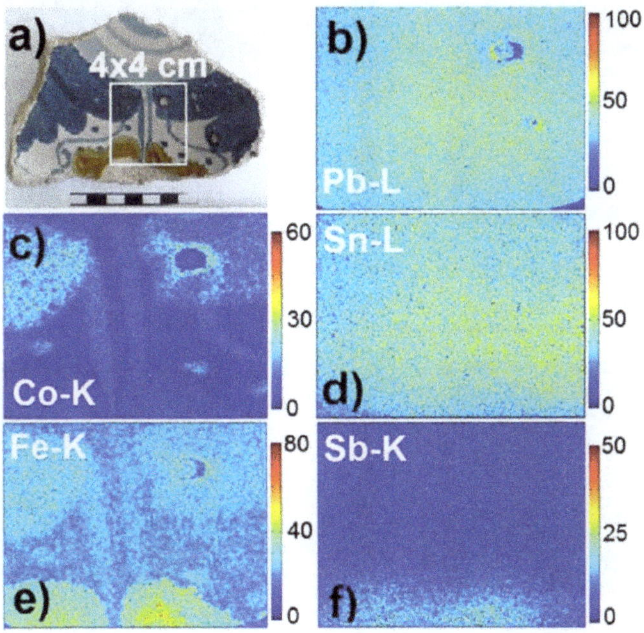

Figure 2.21. XRF imaging of a Nazca painted fragment. (a) A photo of the non-flat sample. The rectangle encloses the area visible in the XRF imaging. (b)–(f) Elemental maps of lead, cobalt, tin, iron, and antimony. Adapted with permission from figure 11 in reference [47], copyright (2014) American Chemical Society.

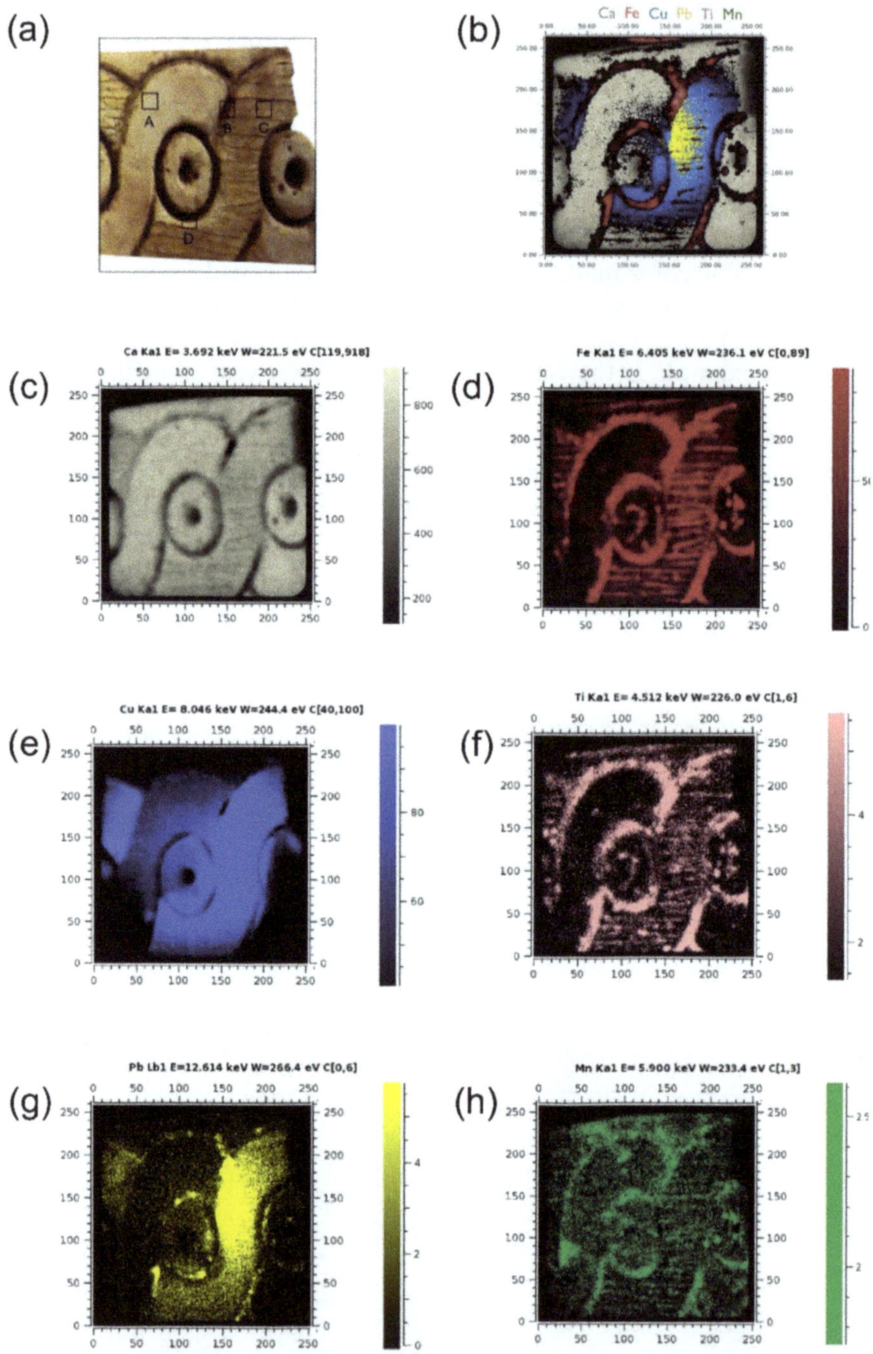

Figure 2.22. XRF imaging of a 2800-year-old Phoenician ivory object. (a) A photo of the sample. Its original painted colors have vanished. (b) False colors inferred from the elemental maps. (c)–(h) Elemental maps of calcium, iron, copper, titanium, lead, and manganese, respectively. Adapted with permission from figures 2, 3, and 4 in reference [49], copyright (2013) American Chemical Society.

figures 2.22(c)–(h). The XRF imaging experiment was carried out at the x-ray fluorescence beamline, ANKA-FLUO, KIT, Germany, utilizing 20 keV monochromatic primary x-rays. This photon energy is much higher than the K-edges of calcium (4.038 keV), titanium (4.965 keV), manganese (6.540 keV), iron (7.112 keV), and copper (8.979 keV), as well as the L_1-edge of lead (15.860 keV). The detector was a pnCCD camera (264×264 pixels, pixel size 48 μm, cooled to −26 °C), which was specially designed for x-ray color imaging; its energy resolution was 152 eV at 5.9 keV. The x-ray optical component was a straight polycapillary optic that was 30.5 mm long and had capillaries 48 μm in diameter; it was capable of projecting 1:1 magnification images. The viewing area was 11.9 mm × 12.3 mm, and the spatial resolution achieved was 48 μm. The total measurement time was less than 100 min. Maps of calcium, iron, copper, titanium, lead, and manganese were generated using the XRF signals of calcium $K\alpha$ (3.692 keV), iron $K\alpha$ (6.405 keV), copper $K\alpha$ (8.046 keV), titanium $K\alpha$ (4.512 keV), lead $L\beta_1$ (12.614 keV), and manganese (5.900 keV), respectively. It was inferred that calcium is the main element of the ivory matrix. Iron corresponds to red ocher (Fe_2O_3). Titanium originates from clay minerals of iron oxides in the form of, for example, ilmenite ($FeTiO_3$). Lead is present either due to a lead-bearing pigment or absorption from the environment. Manganese is present due to the absorption of soil minerals while the object was buried. Finally, false colors were assigned based on the above inferences, as shown in figure 2.22(b).

To date, EPMA has also been applied in the study of the chemical composition of cultural heritage. However, it is crucial to exercise great caution when using EPMA due to the potential radiation damage caused by high-energy electron beams. On the other hand, scanning-type XRF imaging has gained wider popularity than projection-type XRF imaging. Nevertheless, projection-type imaging also offers distinct advantages. It is highly flexible and can be used with uneven samples such as pottery fragments. Furthermore, since it does not require any instrument motion, it reduces the risk of accidentally hitting and damaging valuable cultural heritage samples.

2.6 Chemistry

Chemical reactions such as reaction-diffusion and diffusion-limited aggregation have been extensively studied. These reactions encompass chemical gardens, Liesegang rings, the iodine clock, the Belousov–Zhabotinsky reaction, and so on. People have always been fascinated by their mysterious appearance and the spontaneous formation of regular patterns in gel and liquid environments. Some of these reactions involve growth, maturation, and aging processes, which led to the belief centuries ago that they might be a form of life. Certainly, it is now common knowledge that they cannot be classified as organisms due to the absence of a central dogma or genetic information. Nevertheless, in recent years, numerous studies have linked these phenomena to fractal or fluid dynamics. Investigating the mechanisms underlying these phenomena not only satisfies our innate curiosity but also holds potential for the fabrication of self-assembling ordered material structures. Then, the challenge lies in determining the perspective from which to perform these studies and selecting the appropriate characterization tools.

Studying the diffusion and redistribution of elements can provide valuable insights. In gel and liquid environments, the fate of chemical elements becomes complex. They can be transported to different locations, driven by chemical/electric potential and laminar/turbulent flow, ultimately resulting in changes in local concentration and stoichiometric coefficients. In addition, they may undergo oxidation or reduction, leading to alterations in their valence states. Furthermore, free ions in solutions can mix with one another, and some may combine and precipitate in various forms, either amorphous or crystallized. These changes often manifest as variations in colors and morphologies. However, it is often challenging to accurately identify and distinguish such effects using optical microscopy alone.

To tackle this challenge, it is highly important to develop real-time and *in situ* imaging techniques. For instance, x-ray absorption fine structure (XAFS) imaging can be utilized to elucidate the distribution of valence states, while x-ray diffraction (XRD) imaging can provide valuable insights into crystal structures. Among these techniques, XRF movie imaging has particular importance, as it enables the real-time observation of chemical element diffusion, which is often the primary focus in these studies.

Ion-exchange resins, which have the ability to exchange certain ions present in a solution for other ions of similar charge, have been widely used in a variety of industries, including water treatment, pharmaceuticals, chemical processing, and nuclear power generation. These resins consist of a 3D network of polymer chains to which ionic functional groups are attached. Their performance in ion exchange has been usually only been evaluated by analyzing the changing concentrations of ions in the solution. Therefore, it is extremely interesting to visualize how the ions are captured by the resin during the ion-exchange process. XRF movie imaging is one of the most suitable methods for observing chemical diffusion within ion-exchange resins. For cation exchange resins, the process of metal cations gradually diffusing into resin particles can be recorded by XRF movie imaging. As shown in figure 2.23, ion-exchange resins were placed on a wet filter paper with a copper sulfate solution [14]. The experimental conditions used for the XRF imaging were the same as those of figure 2.2, except that the photon energy of the primary x-rays was set to 9.1 keV, just above the copper K-edge (8.979 keV). In the case of synchrotron experiments, the use of an absorption edge is a convenient method. Even if one is simply recording images without analyzing the energy of the x-rays produced by the sample, the origin of the image can be identified as copper by just checking that the image becomes dark at 8.8 keV, the lower energy side of the copper K-edge, compared to 9.1 keV. The exposure time for each movie image was 1 min. One image was taken using 8.8 keV primary x-rays with a 2 min exposure for the purpose of comparison.

As shown in figure 2.23(A), it is evident that at the beginning of the reaction, copper was evenly distributed in the solution that infiltrated the filter paper. As the reaction progressed, the amount of copper in the solution decreased and copper gradually accumulated in the particles of the ion-exchange resin. In the magnified images of figure 2.23(B), it can be observed that at an early stage, copper only

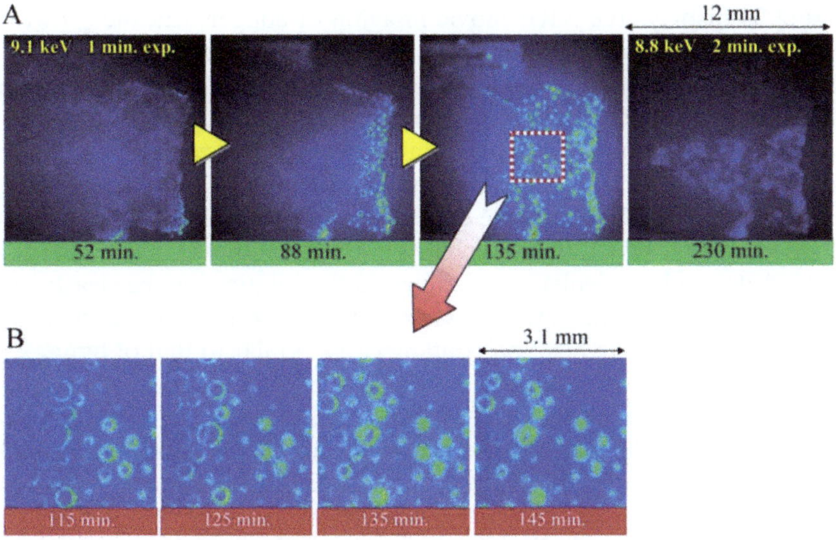

Figure 2.23. The time dependence of XRF images of ion-exchange resins on wet filter paper with a copper sulfate solution. (A) XRF images at the 52nd, 88th, and 135th minutes. The image of scattering x-rays taken with 8.8 keV primary x-rays (below the copper K-absorption edge) is also displayed. (B) Enlarged XRF images at the 115th, 125th, 135th, and 145th minutes. Adapted with permission from figure 3 in reference [14], copyright (2002) American Chemical Society.

permeated the surface layer of the ion-exchange resin, exhibiting a hollow, bubble-like shape. Subsequently, copper entered the interior of the ion-exchange resin, and the 'bubbles' became solid.

Another interesting example is that of electrodeposition, which is a widely used technique for coating, painting, and alloying. During the electrodeposition of some metals, such as lithium, iron, copper, zinc, silver, tin, and lead, variable patterns may form and grow. Under quasi 2D conditions, typical patterns include stringy, finger-like, dendritic, open ramified, mesh-like, dense branching, and diffusion-limited aggregated morphologies [50]. The deposited patterns depend on the growth conditions, including the applied voltage, the concentration of the solution, and the thickness of the electrolyte layer. Studying the electrodeposition process can help us to understand the related diffusion–reaction chemical mechanisms and control the properties of electrodeposited products. In addition, in lithium-ion or lead–acid batteries, lithium or lead may deposit on the cathode in the form of dendritic structures during charge–discharge cycles. The deposition of dendrites is often associated with a deterioration in battery performance and may even cause accidents. To date, optical microscopy has often been used to study the electro-deposition process. X-ray diffraction can also be used to analyze the crystal structures of electrodeposited products. In addition, real-time XRF movie imaging can serve as a powerful tool for tracking the distributions of different chemical elements during the reaction. This is particularly important for binary, ternary, or multicomponent alloy electrodeposition systems.

Figure 2.24 displays the XRF movie imaging of copper dendrite growth during electrodeposition [51]. The reaction took place in a horizontally placed, quasi 2D container. A needle-shaped cathode was positioned in the center of the container, while a metal ring at the container edge served as the cathode. The solution used for the reaction was 0.1 mol L^{-1} $CuSO_4$, and the applied voltage was 2.8 V. The experimental conditions were the same as those of figure 2.13, except that the photon energy was set to 9.2 keV, above the copper K-edge (8.979 keV). The exposure time for each movie image was 100 milliseconds. In the XRF movie frames, the growth of copper dendrites and the depletion of copper in the solution can be clearly observed.

Figure 2.25 displays the XRF movie imaging of zinc dendrite growth during electrodeposition [51]. The reaction container was similar to that of figure 2.24. The solution used for the reaction was 0.36 mol L^{-1} $ZnSO_4$, and the applied voltage was 2.5 V. The experimental conditions were the same as those of figure 2.24, except that the photon energy of the primary x-rays was set to 9.8 keV, above the zinc K-edge (9.659 keV). In the XRF movie frames, the growth of zinc dendrites can be clearly observed.

Figure 2.26 presents a similar demonstration, showing the deposition of zinc dendrites from a 0.2 mol L^{-1} zinc chloride solution on a copper cathode [52]. The applied voltage was 3 V. The XRF movie imaging experiment was carried out at

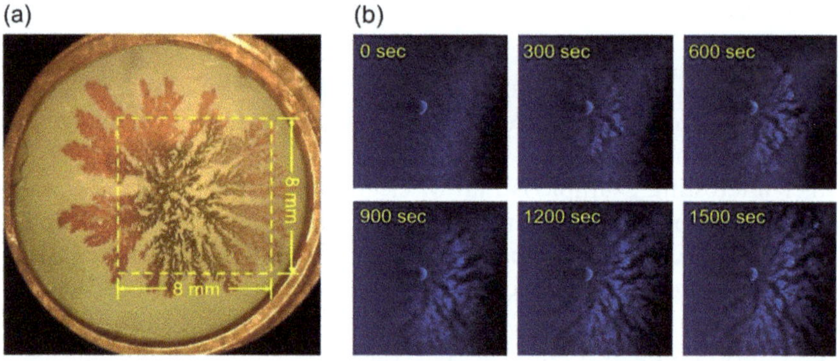

Figure 2.24. XRF movie imaging of copper dendrite growth during electrodeposition. (a) An optical photomicrograph of the dendrites after the reaction. (b) XRF movie frames captured at different stages of the reaction [51].

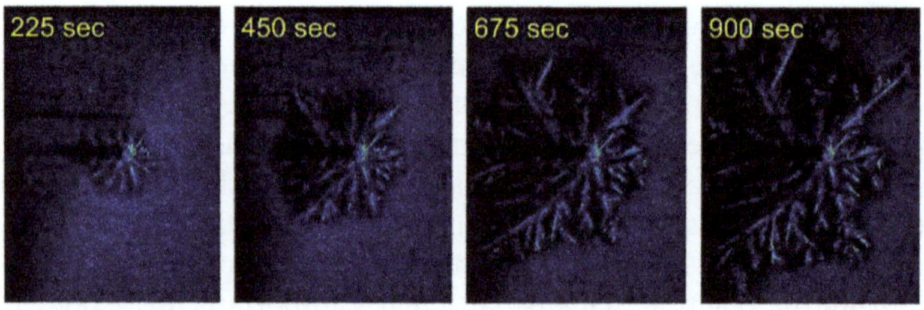

Figure 2.25. XRF movie imaging of zinc dendrite growth during electrodeposition [51].

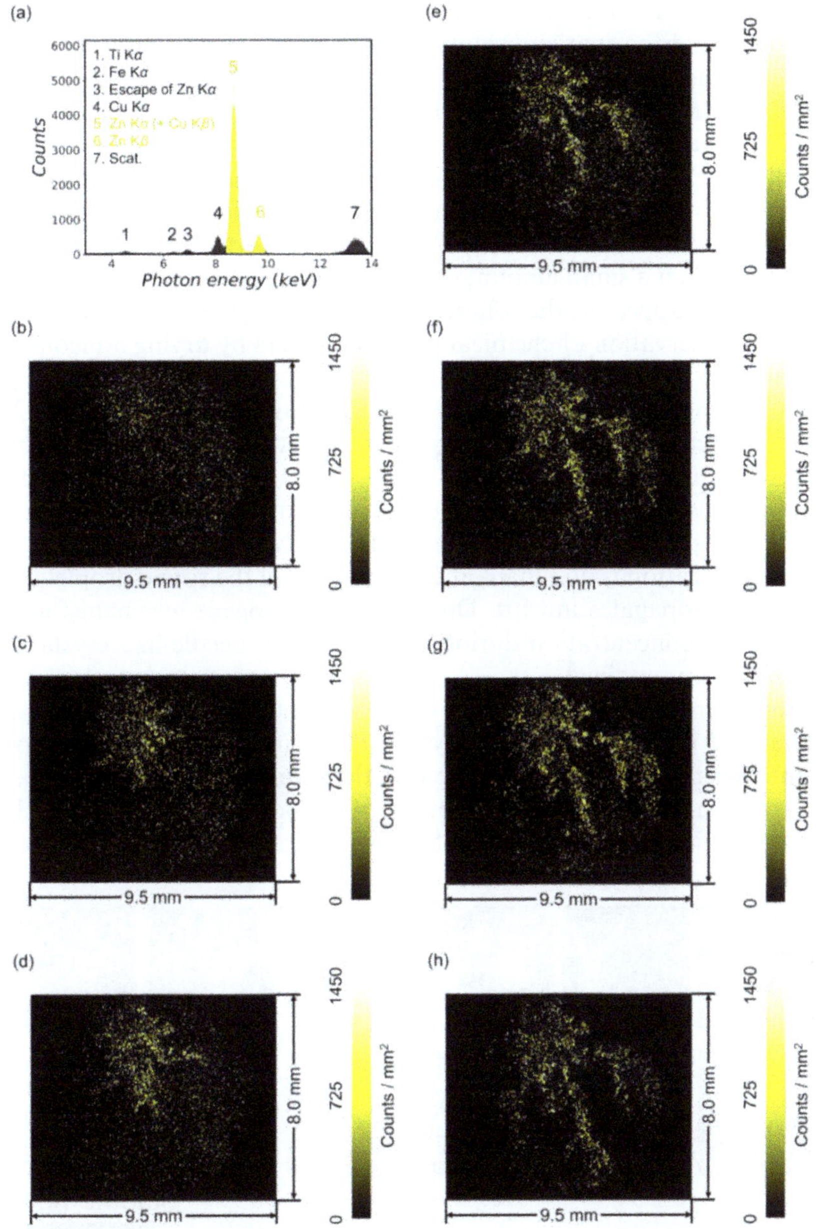

Figure 2.26. XRF movie imaging of zinc dendrite growth during electrodeposition. (a) The XRF spectrum of the chemical garden. (b)–(h) Movie frames of zinc (Zn Kα and Kβ) at 0, 4, 8, 12, 16, 20, and 58 min, respectively. The acquisition time of each movie frame was two minutes. Modified from figure 2 in reference [52] CC BY 2.0.

BL-14B, Photon Factory, KEK. Japan, utilizing 13.5 keV monochromatic primary x-rays, the photon energy of which is higher than the K-edges of copper (8.979 keV) and zinc (9.659 keV). The detector was the same visible-light digital scientific camera as that utilized for figure 2.14. The optical component was a pinhole 50 μm in diameter manufactured on a 50 μm thick tungsten foil. The imaging magnification was 1.75 times, and the spatial resolution achieved was estimated to be 80 μm. The measurement time for each movie frame was 2 min. In the XRF spectra accumulated during the reaction, prominent zinc peaks were clearly observed, along with other spectral peaks from the polymer reaction container and copper electrodes. Despite the limited intensity in the movie frames, we could still clearly observe the gradual growth of zinc dendrites and the depletion of zinc in the solution.

It is well known that some chemical patterns are formed during crystal growth from solution. When a small amount of solution is dropped onto the substrate, such patterns gradually appear as the solution evaporates. Figures 2.27 and 2.28 show the x-ray imaging observation of chemical patterns formed by drying aqueous solutions of potassium ferrocyanide, $K_4[Fe(CN)_6]$, and potassium bromide, KBr, respectively [24]. The experiments were performed at the undulator beamline BL-NW2A1, PF-AR (Photon Factory's Advanced Ring operated at 6.5 GeV), KEK, Japan. The detector used was a CCD camera (TC281 CCD, made by Texas Instruments, 1000 × 1000 pixels, pixel size 8 μm, cooled to −30 °C). The incident x-ray energy was set to 7.4 keV (higher than the Fe K-edge and the K K-edge). In figure 2.27, it can be seen that crystallization is dominant at the periphery of the round droplet at an early stage and later propagates inward. Due to the inhomogeneous change in chemical composition and concentration during crystallization, needle-like crystals of quite different sizes were obtained.

Potassium ferrocyanide ($K_4[Fe(CN)_6]$) in aqueous solution not only provides unique chemical patterns but also plays a role in another scientifically interesting experiment. When it comes into contact with the surface of crystalline copper sulfate ($CuSO_4$) grains, a reddish-brown copper ferrocyanide ($Cu_2[Fe(CN)_6]$) film begins to form at the solid–liquid interface. It grows in a bag-like shape and has an

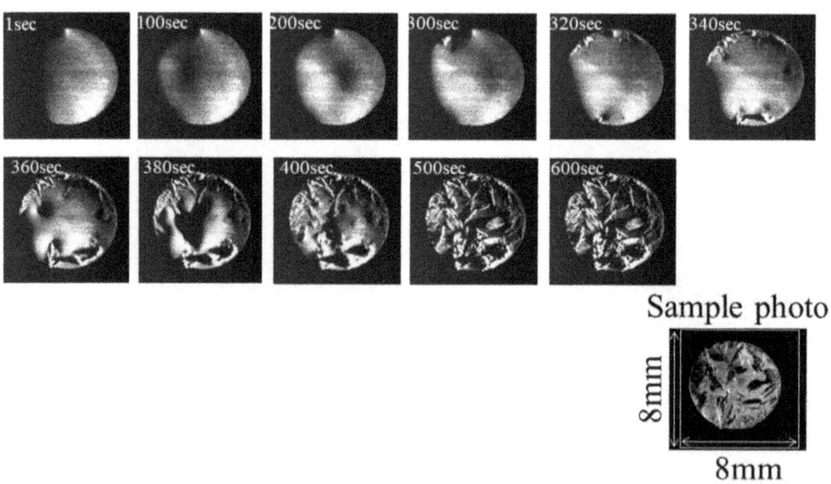

Figure 2.27. Real-time observation of chemical pattern formation in crystalline $K_4[Fe(CN)_6]$ grown from an aqueous solution. The incident x-ray energy was 7.4 keV. The exposure time was 1 s/frame. An optical microscope image of the crystallized residue sample is also shown [24].

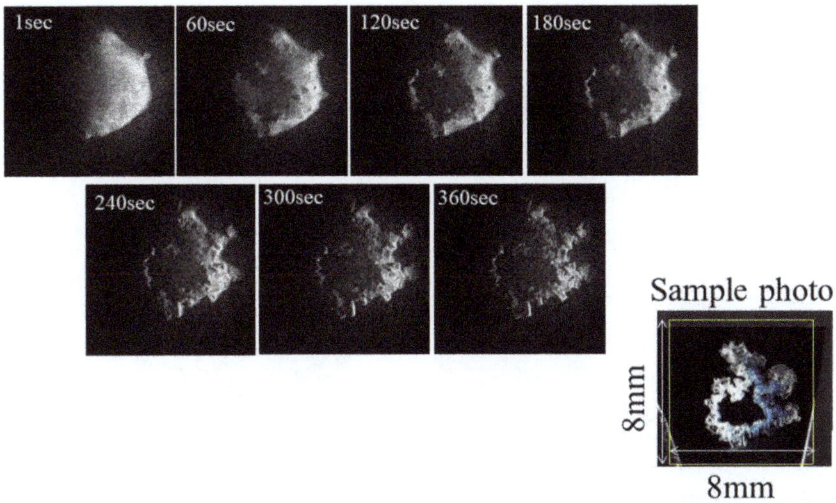

Figure 2.28. Real-time observation of chemical pattern formation in crystalline KBr grown from an aqueous solution. The incident x-ray energy was 7.4 keV. The exposure time was 1 s/frame. An optical microscope image of the crystallized residue sample is also shown [24].

appearance similar to a biological cell, known as 'Traube's artificial cell,' since it was discovered by Moritz Traube (1826–1894) [53]. Figure 2.29(a) shows a series of optical microscope images obtained during the changes that take place following contact with $CuSO_4$ grains [24]. The reddish-brown film forms over almost the entire surface of the grain within 30 s, extending and growing toward the periphery (liquid part). The final size of the film can reach 1–4 mm. The film growth is not continuous but proceeds with a marked change in shape, such as sudden expansion and local deformation. Initially, the change is rapid, lasting less than 1 sec, but it eventually becomes slow. Figures 2.29(b) and (c) show x-ray images obtained using different incident x-ray energies, namely 9.3 keV (higher than the Cu K-edge) and 7.4 keV (higher than the Fe K-edge but lower than the Cu K-edge), respectively [24]. It can be seen that copper is found only in the red-brown membrane area and not outside it. This corresponds to the fact that the membrane is semipermeable.

2.7 Simultaneous multielement movies

The application of XRF movie imaing, which is time-resolved x-ray color imaging, can be even more powerful when the reaction involves a complicated inhomogeneous multielement system. The chemical garden is a well-known example of such a phenomenon. The typical reaction is initiated by seeding a soluble metal salt crystal in an aqueous solution of sodium silicate, Na_2SiO_3. As the crystal begins to dissolve, it is immediately surrounded by a semipermeable membrane that precipitates at the interface between the internal metal salt solution and the external silicate solution. The chemical garden was first observed and described by Johann Rudolf Glauber in 1646 [54]. The name 'chemical garden' is derived from the fact that it resembles the

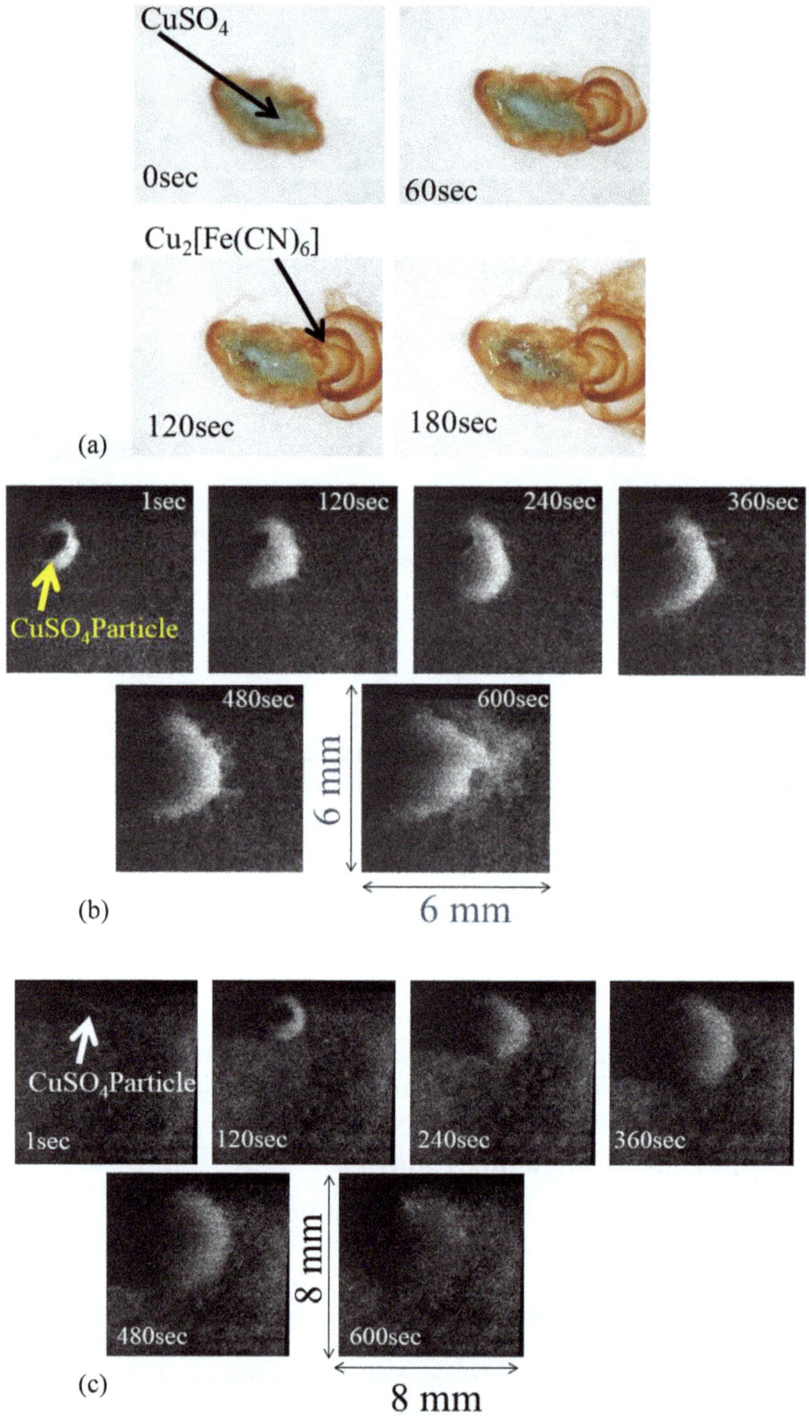

Figure 2.29. Real-time observation of Traube's artificial cell. (a) Optical microscope observation. (b) X-ray observation with an incident energy of 9.3 keV (higher than the Cu K-edge), (c) X-ray observation with an incident energy of 7.4 keV (higher than the Fe K-edge but lower than the Cu K-edge) [24].

growth of natural plants. The simplified explanation of the mechanism is as follows [55–59]: when the particle of the metal salt crystal begins to dissolve, a semi-permeable membrane shell is immediately precipitated at the interface between the metal salt solution and the silicate solution. The continuous dissolution of the metal salt increases the osmotic pressure inside the shell, causing external water to enter the shell through the semipermeable membrane. As a result, the shell expands and ruptures, and fluid from the metal salt solution is expelled into the external sodium silicate solution. Again, a new shell is formed at the interface of the two solutions, trapping the fluid. This precipitation-rupture process can be repeated many times. Eventually, a plant-like tubular structure emerges and grows. This knowledge is sometimes useful in distinguishing between biological fossils and non-biological structures in ancient rocks. Non-biological chemical gardens have been observed on the surface of Mars [60]. In such cases, the colors of the chemical elements revealed by x-ray fluorescence are clearly powerful because they confirm the presence of non-biological metals.

Figure 2.30 displays the process of a chemical garden reaction recorded by an optical video [61]. The seed, exhibiting a pale yellow color, was a homogeneous mixture of calcium chloride ($CaCl_2$) and ferrous sulfate heptahydrate ($FeSO_4 \cdot 7H_2O$) with equivalent mass. The aqueous sodium silicate solution was obtained by diluting a commercial solution of 55 wt% Na_2SiO_3 with twice the mass of water. The reaction took place in a vertical thin container measuring 30 mm (width) × 15 mm (height) × 0.5 mm (thickness), with a 50 μm thick polyester thin film as the front wall to allow x-rays to penetrate with acceptable attenuation. This purpose of the container design was to facilitate XRF imaging.

The video frames clearly show tubular structures sprouting and growing upward during the reaction. There is also a color segregation of white and green, with the green structures growing slower than the white structures. The color segregation is

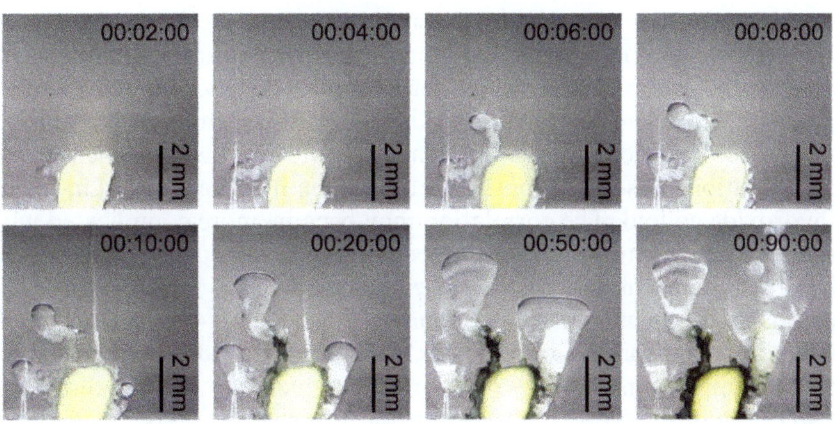

Figure 2.30. A chemical garden reaction captured by optical video, with a timestamp included in each frame. Reproduced with permission from [61].

expected to be related to the different compositions of the precipitated materials. However, inferring the chemical compositions from the optical video alone is difficult.

A similar chemical garden reaction was recorded as an XRF movie [62], as shown in figure 2.31. The experimental conditions for XRF imaging were the same as those of figure 2.10, except that the pinhole diameter was 50 μm, the imaging magnification was 2 times, and the spatial resolution achieved was estimated to be better than 100 μm. The reaction took 15 h, as did the XRF movie recording. Figure 2.31(a) shows an optical photomicrograph of the reaction products after 15 h. It has a similar appearance to that of figure 2.30, including the big swell at the initial position of the seed, tubular structures, and a clear segregation between white and green structures. Figure 2.31(b) shows the XRF spectrum accumulated during the 15 h, where the spectral peaks of calcium and iron are clearly observed. The peaks of calcium are much weaker than those of iron primarily because calcium fluorescence x-rays are more intensively absorbed in the solution. The XRF movie frames of calcium and iron were collected at a frame rate of one frame per hour to adapt to the gradual reaction process. Figures 2.31(c) and (d) show the frames captured for calcium in the 1st and 15th hours, respectively, and figures 2.31(e)–(h) show the frames captured for iron in the 1st, 3rd, 5th, and 15th hours, respectively. The correlation between the optical appearance and the element composition can easily be clarified by comparing the optical photo and the XRF movie frames: the green structures contain both calcium and iron, whereas the white structures contain only calcium. The gradual diffusion of iron from the internal solutions to the precipitated boundary of the big swell is also clear. In contrast to the spatial distribution of iron, the spatial distribution of calcium remains stable over many hours. The reason that iron and calcium exhibit different behaviors in terms of spatial distribution and gradual diffusion is that their hydroxides have very different solubility product constants (5×10^{-6} for calcium hydroxide and 8×10^{-16} for ferrous hydroxide). It should be noted that in this experiment, the correlation between colors and elements is intuitive: white corresponds to calcium and green corresponds to ferrous structures. However, in many reactions, the correlation is not so simple. According to the authors' experience, for example, the precipitated membrane of iron appears green when the ferrous oxidation state is dominant but turns brown when the ferric oxidation state is dominant; cobalt appears red when it takes the form of free cations in solution but turns blue when it is precipitated in the form of cobalt hydroxide colloid. The correlation may become more complicated when two or more metal cations are mixed or coprecipitated. Consequently, a movie imaging technique that can accurately identify chemical elements is indispensable.

As emphasized in section 2.1, in comparison to scanning-type XRF imaging, the greatest advantage of projection-type XRF imaging is its ability to perform movie imaging of samples undergoing reactions, in which the spatial distribution of elements continuously evolves. This extension of projection-type XRF imaging to movie imaging will significantly enhance its applications in the field of chemistry and other related disciplines.

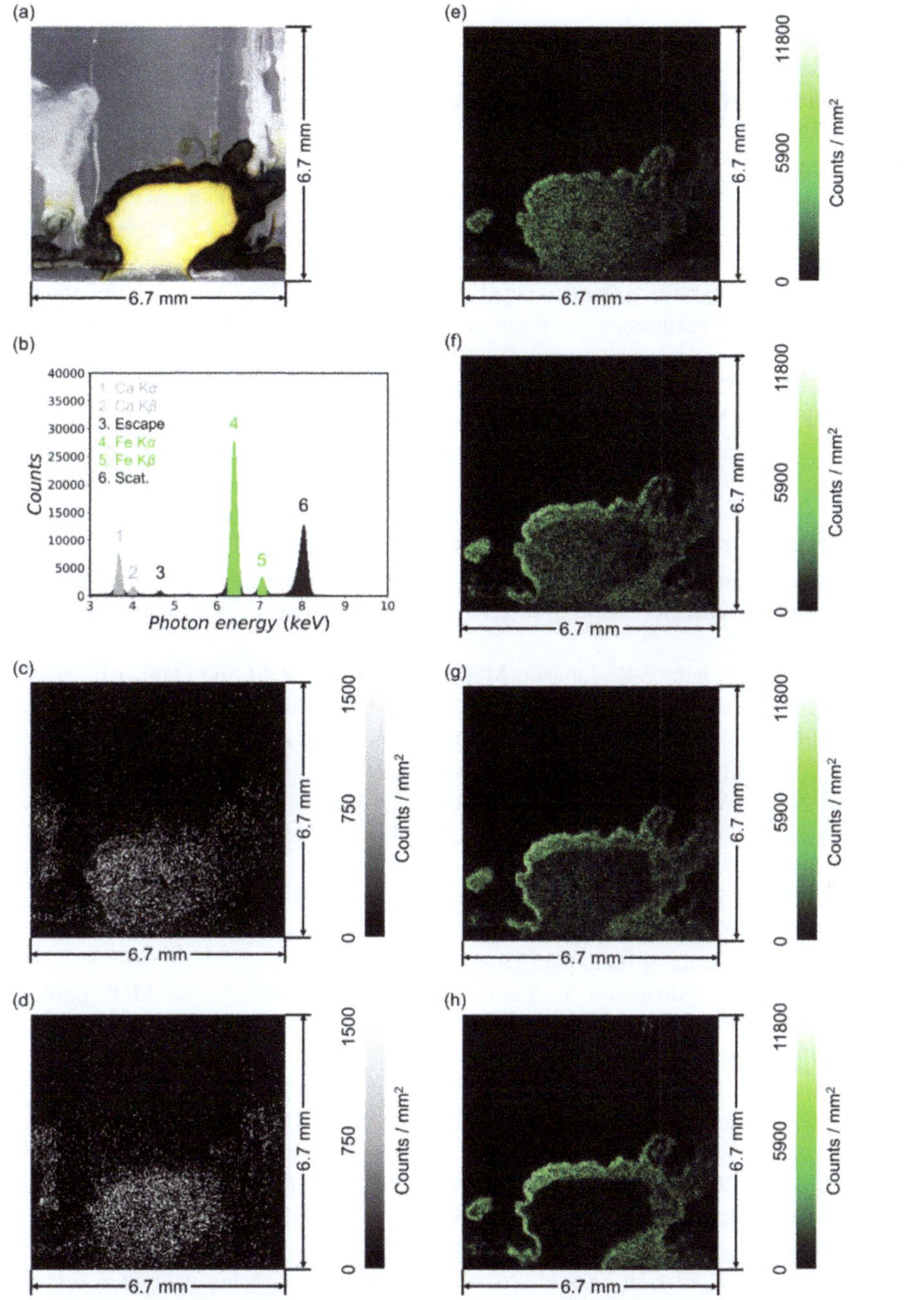

Figure 2.31. XRF movie imaging of a chemical garden reaction. (a) An optical photomicrograph of the final structure after 15 h of the reaction. (b) The XRF spectrum of the chemical garden accumulated over the 15 h. (c) and (d) Movie frames of calcium (Ca Kα and Kβ) in the 1st and 15th h, respectively. (e)–(h) Movie frames of iron (Fe Kα and Kβ) in the 1st, 3rd, 5th, and 15th h, respectively. Modified with permission from figures in reference [62], copyright (2017) American Chemical Society.

References

[1] Sparks C J 1980 X-ray fluorescence microprobe for chemical analysis *Synchrotron Radiation Research* ed H Winick and S Doniach (Boston, MA: Springer) https://doi.org/10.1007/978-1-4615-7998-4

[2] Kemner K M, Kelly S D, Lai B, Maser J, O'Loughlin E J, Sholto-Douglas D, Cai Z, Schneegurt M A, Kulpa C F and Nealson K H 2004 Elemental and redox analysis of single bacterial cells by x-ray microbeam analysis *Science* **306** 686–7 https://doi.org/10.1126/science.1103524

[3] Shimura M, Szyrwiel L, Matsuyama S and Yamauchi K 2017 Visualization of intracellular elements using scanning x-ray fluorescence microscopy In *Metallomics* (Tokyo: Springer) pp 63–92 https://doi.org/10.1007/978-4-431-56463-8_3

[4] Victor T W, Easthon L M, Ge M, O'Toole K H, Smith R J, Huang X, Yan H, Allen K N, Chu Y S and Miller L M 2018 X-ray fluorescence nanotomography of single bacteria with a sub-15 nm beam *Sci Rep.* **8** 13415 https://doi.org/10.1038/s41598-018-31461-y

[5] Hosokawa Y, Ozawa S, Nakazawa H and Nakayama Y 1997 An x-ray guide tube and a desk-top scanning x-ray analytical microscope *X-Ray Spectrom.* **26** 380–7 https://doi.org/10.1002/(SICI)1097-4539(199711/12)26:6<380::AID-XRS237>3.0.CO;2-%23

[6] Haschke M 2014 *Laboratory Micro-X-Ray Fluorescence Spectroscopy* **vol 55** (Springer Series in Surface Sciences) (Cham: Springer International Publishing) https://doi.org/10.1007/978-3-319-04864-2

[7] Eldesoky A, Logan E R, Johnson M B, McFarlane C R M and Dahn J R 2020 Scanning micro x-ray fluorescence (MXRF) as an effective tool in quantifying Fe dissolution in $LiFePO_4$ cells: towards a mechanistic understanding of Fe dissolution *J. Electrochem. Soc.* **167** 130539 https://doi.org/10.1149/1945-7111/abba62

[8] Born M and Wolf E 2020 *Principles of Optics: 60th Anniversary Edition* (London: Cambridge University Press) https://doi.org/10.1017/9781108769914

[9] Jacobsen C 2019 *X-Ray Microscopy* (Cambridge: Cambridge University Press) https://doi.org/10.1017/9781139924542

[10] Boyle W S and Smith G E 1970 Charge coupled semiconductor devices *Bell Syst. Tech. J.* **49** 587–93 https://doi.org/10.1002/j.1538-7305.1970.tb01790.x

[11] Matsumoto K, Nakamura T, Yusa A and Nagai S 1985 A new MOS phototransistor operating in a non-destructive readout mode *Jpn. J. Appl. Phys.* **24** L323 https://doi.org/10.1143/JJAP.24.L323

[12] Paunesku T, Vogt S, Maser J, Lai B and Woloschak G 2006 X-ray fluorescence microprobe imaging in biology and medicine *J. Cell. Biochem.* **99** 1489–502 https://doi.org/10.1002/jcb.21047

[13] Fahrni C J 2007 Biological applications of x-ray fluorescence microscopy: exploring the subcellular topography and speciation of transition metals *Curr. Opin. Chem. Biol.* **11** 121–7 https://doi.org/10.1016/j.cbpa.2007.02.039

[14] Sakurai K and Eba H 2003 Micro x-ray fluorescence imaging without scans: toward an element-selective movie *Anal. Chem.* **75** 355–9 https://doi.org/10.1021/ac025793h

[15] Hogarth P 2015 Neurodegeneration with brain iron accumulation: diagnosis and management *J. Mov. Disord.* **8** 1–13 https://doi.org/10.14802/jmd.14034

[16] Kirilina E *et al* 2020 Superficial white matter imaging: contrast mechanisms and whole-brain *in vivo* mapping *Sci. Adv.* **6** eaaz9281 https://doi.org/10.1126/sciadv.aaz9281

[17] Skorbiłowicz M, Skorbiłowicz E and Cieśluk I 2018 Bees as bioindicators of environmental pollution with metals in an urban area *J. Ecol. Eng.* **19** 229–34 https://doi.org/10.12911/22998993/85738

[18] Kump P, Nečemer M and Šnajder J 1996 Determination of trace elements in bee honey, pollen and tissue by total reflection and radioisotope x-ray fluorescence spectrometry *Spectrochim. Acta Part B At. Spectrosc.* **51** 499–507 https://doi.org/10.1016/0584-8547(95)01435-7

[19] Romano F P, Caliri C, Cosentino L, Gammino S, Mascali D, Pappalardo L, Rizzo F, Scharf O and Santos H C 2016 Micro x-ray fluorescence imaging in a tabletop full field-x-ray fluorescence instrument and in a full field-particle induced x-ray emission end station *Anal. Chem.* **88** 9873–80 https://doi.org/10.1021/acs.analchem.6b02811

[20] Kühn R, Pattard M, Pernak K-D and Winter A 1989 Results of the harmful effects of water pollutants to Daphnia Magna in the 21 day reproduction test *Water Res.* **23** 501–10 https://doi.org/10.1016/0043-1354(89)90142-5

[21] Abdullahi M, Li X, Abdallah M A-E, Stubbings W, Yan N, Barnard M, Guo L-H, Colbourne J K and Orsini L 2022 Daphnia as a sentinel species for environmental health protection: a perspective on biomonitoring and bioremediation of chemical pollution *Environ. Sci. Technol.* **56** 14237–48 https://doi.org/10.1021/acs.est.2c01799

[22] De Samber B *et al* 2019 Three-dimensional x-ray fluorescence imaging modes for biological specimens using a full-field energy dispersive CCD camera *J. Anal. At. Spectrom.* **34** 2083–93 https://doi.org/10.1039/c9ja00198k

[23] Pena L D, Cacho I, Calvo E, Pelejero C, Eggins S and Sadekov A 2008 Characterization of contaminant phases in foraminifera carbonates by electron microprobe mapping *Geochem. Geophys. Geosyst.* **9** n/a-n/a https://doi.org/10.1029/2008GC002018

[24] Sakurai K and Mizusawa M unpublished data

[25] Wang X and Guo Y 2021 The impact of trace metal cations and absorbed water on colour transition of turquoise *R. Soc. Open Sci.* **8** 201110 https://doi.org/10.1098/rsos.201110

[26] Abe Y, Nakamura A, Suzuki S, Tantrakarn K, Nakai I, Zöldföldi J and Pfälzner P 2019 Use of variscite as a gemstone in the late bronze age Royal Tomb at Qatna, Syria *J. Archaeol. Sci. Rep.* **27** 101994 https://doi.org/10.1016/j.jasrep.2019.101994

[27] Fisher L A *et al* 2015 Quantified, multi-scale x-ray fluorescence element mapping using the maia detector array: application to mineral deposit studies *Miner. Depos.* **50** 665–74 https://doi.org/10.1007/s00126-014-0562-z

[28] Devès G, Perroux A S, Bacquart T, Plaisir C, Rose J, Jaillet S, Ghaleb B, Ortega R and Maire R 2012 Chemical element imaging for speleothem geochemistry: application to a uranium-bearing corallite with aragonite diagenesis to opal (Eastern Siberia, Russia) *Chem. Geol.* **294–295** 190–202 https://doi.org/10.1016/j.chemgeo.2011.12.003

[29] Zhao W and Sakurai K 2017 CCD camera as feasible large-area-size x-ray detector for x-ray fluorescence spectroscopy and imaging *Rev. Sci. Instrum.* **88** 063703 https://doi.org/10.1063/1.4985149

[30] Chen P, Chen T, Xu L, Liu H and Xie Q 2017 Mn-rich limonite from the yeshan iron deposit, Tongling District, China: a natural nanocomposite *J. Nanosci. Nanotechnol.* **17** 6931–5 https://doi.org/10.1166/jnn.2017.14410

[31] Zhao W, Hirano K and Sakurai K 2019 Expanding a polarized synchrotron beam for full-field x-ray fluorescence imaging *Rev. Sci. Instrum.* **90** 113704 https://doi.org/10.1063/1.5115421

[32] Mizusawa M and Sakurai K 2004 XAFS imaging of tsukuba gabbroic rocks: area analysis of chemical composition and local structure *J. Synchrotron Radiat.* **11** 209–13 https://doi.org/10.1107/S0909049503028024

[33] Re A, Giudice A L, Angelici D, Calusi S, Giuntini L, Massi M and Pratesi G 2011 Lapis lazuli provenance study by means of micro-PIXE *Nucl. Instrum. Methods Phys. Res. Sect. B Beam Interact. Mater. Atoms* **269** 2373–7 https://doi.org/10.1016/j.nimb.2011.02.070

[34] Sakurai K and Mizusawa M 2004 Fast x-ray fluorescence camera combined with wide band pass monochromatic synchrotron beam *AIP Conf. Proc. 705* (Synchrotron Radiation Instrumentation 2003) *(San Francisco)* p 889 https://doi.org/10.1063/1.1757938

[35] Zhao W and Sakurai K unpublished data

[36] Matsuyama S, Yamada J, Kohmura Y, Yabashi M, Ishikawa T and Yamauchi K 2019 Full-field x-ray fluorescence microscope based on total-reflection advanced Kirkpatrick–Baez mirror optics *Opt. Express* **27** 18318 https://doi.org/10.1364/OE.27.018318

[37] Feller R L (ed) 1986 *Artists' Pigments: A Handbook of Their History and Characteristics* vol 1 (London: Archetype Publications) https://www.nga.gov/research/publications/pdf-library/artists-pigments-vol-1.html

[38] Roy A (ed) 1993 *Artists' Pigments: A Handbook of Their History and Characteristics* vol 2 (London: Archetype Publications) https://www.nga.gov/research/publications/pdf-library/artists-pigments-vol2.html

[39] FitzHugh E W (ed) 1997 *Artists' Pigments: A Handbook of Their History and Characteristics* vol 3 (London: Archetype Publications) https://www.nga.gov/research/publications/pdf-library/artists-pigments-vol-3.html

[40] Berrie B H (ed) 2007 *Artists' Pigments: A Handbook of Their History and Characteristics* vol 4 (London: Archetype Publications) https://archetype.co.uk/our-titles/artists-pigments-4/?id=35

[41] Gliozzo E and Ionescu C 2022 Pigments—lead-based whites, reds, yellows and oranges and their alteration phases *Archaeol. Anthropol. Sci.* **14** 17 https://doi.org/10.1007/s12520-021-01407-z

[42] Chappell J 1991 *The Potter's Complete Book of Clay and Glazes* Revised (New York: Watson-Guptill)

[43] Appel K, Dik J, Janssens K, Snickt G, Van Der Loeff, Van Der L and Rickers K 2008 Visualization of lost painting by Vincent van Gogh using synchrotron radiation based x-ray fluorescence elemental mapping *Anal. Chem.* **80** 6436–42 https://doi.org/10.1021/ac800965g

[44] Janssens K, Alfeld M, Van Der Snickt G, De Nolf W, Vanmeert F, Radepont M, Monico L, Dik J, Cotte M, Falkenberg G *et al* 2013 The use of synchrotron radiation for the characterization of artists' pigments and paintings *Annu. Rev. Anal. Chem.* **6** 399–425 https://doi.org/10.1146/annurev-anchem-062012-092702

[45] Alfeld M and de Viguerie L 2017 Recent developments in spectroscopic imaging techniques for historical paintings—a review *Spectrochim. Acta— B At. Spectrosc.* **136** 81–105 https://doi.org/10.1016/j.sab.2017.08.003

[46] Walter P, Sarrazin P, Gailhanou M, Hérouard D, Verney A and Blake D 2019 Full-field XRF instrument for cultural heritage: Application to the study of a Caillebotte painting *X-Ray Spectrom.* **48** 274–81 https://doi.org/10.1002/xrs.2841

[47] Romano F P, Caliri C, Cosentino L, Gammino S, Giuntini L, Mascali D, Neri L, Pappalardo L, Rizzo F, Taccetti F *et al* 2014 Macro and micro full field x-ray fluorescence with an x-ray pinhole camera presenting high energy and high spatial resolution *Anal. Chem.* **86** 10892–9 https://doi.org/10.1021/ac503263h

[48] Schweizer F and Rinuy A 1982 Manganese black as an etruscan pigment *Stud. Conserv.* **27** 118–23 https://doi.org/10.1179/sic.1982.27.3.118

[49] Reiche I, Müller K, Albéric M, Scharf O, Wähning A, Bjeoumikhov A, Radtke M and Simon R 2013 Discovering vanished paints and naturally formed gold nanoparticles on 2800 years old Phoenician ivories using SR-FF-MicroXRF with the color x-ray camera *Anal. Chem.* **85** 5857–66 https://doi.org/10.1021/ac4006167

[50] Eba H and Sakurai K 2004 Pattern transition in Cu-Zn binary electrochemical deposition *J. Electroanal. Chem.* **571** 149–58 https://doi.org/10.1016/j.jelechem.2004.05.024

[51] Eba H and Sakurai K unpublished data

[52] Zhao W and Sakurai K 2019 Multi-element x-ray movie imaging with a visible-light CMOS camera *J. Synchrotron Radiat.* **26** 230–3 https://doi.org/10.1107/S1600577518014273

[53] Traube M 1867 Experimente zur Theorie der Zellenbildung und Endosmose *Arch. Anat. Physiol. wiss. Med.* 87–165

[54] Glauber J R 1646 Wie man in diesem Liquore von allen Metallen in wenig Stunden Bäume mit Farben soll wachsen machen [How one shall make grow—in this solution, from all metals, in a few hours—trees with color] (in German) *Furni Novi Philosophici* 2nd edn (Amsterdam: Johan Jansson) pp 157–60 https://digitale-sammlungen.gwlb.de/resolve?PPN=862388155

[55] Cartwright J H E, García-Ruiz J M, Novella M L and Otálora F 2002 Formation of chemical gardens *J. Colloid Interface Sci.* **256** 351–9 https://doi.org/10.1006/jcis.2002.8620

[56] Cartwright J H E, Escribano B and Sainz-Díaz C I 2011 Chemical-garden formation, morphology, and composition. I. Effect of the nature of the cations *Langmuir* **27** 3286–93 https://doi.org/10.1021/la104192y

[57] Cartwright J H E, Escribano B, Sainz-Díaz C I and Stodieck L S 2011 Chemical-garden formation, morphology, and composition. II. Chemical gardens in microgravity *Langmuir* **27** 3294–300 https://doi.org/10.1021/la104193q

[58] Barge L M *et al* 2015 From chemical gardens to chemobrionics *Chem. Rev.* **115** 8652–703 https://doi.org/10.1021/acs.chemrev.5b00014

[59] Cardoso S S S *et al* 2020 Chemobrionics: from self-assembled material architectures to the origin of life *Artif. Life* **26** 315–26 https://doi.org/10.1162/artl_a_00323

[60] Sainz-Díaz C I, Escribano B, Sánchez-Almazo I and Cartwright J H E 2021 Chemical gardens under mars conditions: imaging chemical garden growth *in situ* in an environmental scanning electron microscope *Geophys. Res. Lett.* **48** e2021GL092883 https://doi.org/10.1029/2021GL092883

[61] Sakurai K and Zhao W 2018 Novel x-ray technique to visualize moving elements *Tech. Assoc. Refractories, Japan* **70** 399–403 (in Japanese)

[62] Zhao W and Sakurai K 2017 Realtime observation of diffusing elements in a chemical garden *ACS Omega* **2** 4363–9 https://doi.org/10.1021/acsomega.7b00930

IOP Publishing

X-ray Color Imaging
Static and dynamic x-ray fluorescence for chemical element identification
Kenji Sakurai and Wenyang Zhao

Chapter 3

Methods and instruments for x-ray color imaging

3.1 Compatibility with conventional x-ray fluorescence spectrometry

As explained earlier, there are two different ways to image chemical elements using x-rays. Figure 3.1 schematically compares the concepts underlying the scanning and projection types of 2D x-ray fluorescence (XRF) imaging of the chemical elements in a sample.

From an instrumental standpoint, scanning-type XRF imaging is very similar to conventional XRF spectrometry. The main difference is the use of a microbeam. With the addition of x-ray focusing optics and a sample scanning stage, imaging can be performed. This technique provides XRF spectra at specific points during the x–y scanning of the sample. At the end of the scanning process, scanning-type XRF imaging can also give the average chemical composition, which is equivalent to the output of ordinary wide-beam XRF analysis.

Projection-type XRF imaging always obtains information for the whole analysis area of the sample, like ordinary wide-beam XRF analysis. The main difference between them is the use of a 2D x-ray detector instead of a conventional energy-dispersive x-ray detector, such as a silicon drift detector (SDD) and/or Si(Li) detector. The imaging optics must also be installed between the sample and the 2D detector. In principle, the technique always obtains x-ray signals for all pixels corresponding to all points in the sample. Therefore, it is possible to obtain an image of specific chemical elements and at the same time the average chemical composition of the whole sample. In other words, the use of projection-type XRF imaging can retain backward compatibility with ordinary XRF spectrometry.

Another interesting comparison concerns the number of images to be processed. In the case of the scanning method, the image corresponding to a specific chemical element is obtained at the end of scanning. Therefore, one image must be handled for each chemical element. On the other hand, in the case of the projection

doi:10.1088/978-0-7503-3215-6ch3

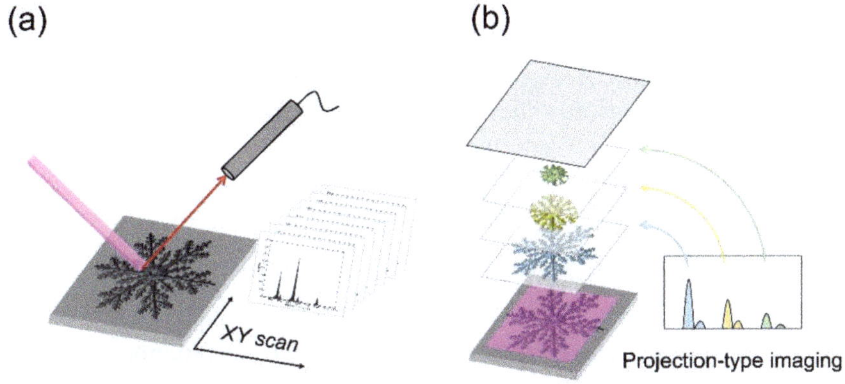

Figure 3.1. Two major methods used in x-ray color imaging. (a) The scanning method and (b) the projection method.

Table 3.1. A comparison of instruments used for scanning-type and projection-type XRF imaging.

	Scanning-type	Projection-type
Primary x-ray beam	Micro or nano focused	Wide
Image transfer optics	—	Required
Detector	Energy resolution, e.g. SDD	Energy resolution and pixel resolution e.g. energy x-ray imagers such as p–n charge-coupled devices (CCDs). CCDs and complementary metal–oxide–semiconductor (CMOS) cameras are used in special operations.
Focus of primary x-ray beam	Required	—
Scanning system	Required	—
Resolution	10–100 μm in the laboratory, up to 30 nm in synchrotron	Mostly 10–100 μm; an exceptional example has reached 500 nm
Available for time-dependent samples	No	Yes

approach, the images are taken continuously over a short time. The number of images to be processed can sometimes be huge. Using image processing, a single image is finally obtained for each chemical element.

Although both types of imaging give essentially the same information when the imaging process is complete, projection-type imaging has some competitive advantages and holds great potential for widespread future application. Table 3.1 briefly summarizes these points.

To achieve projection-type XRF imaging, some devices are necessary. Among the most important parts are the x-ray optical component and the 2D x-ray energy-dispersive detector. The detector should be able to simultaneously record x-ray photon energy and position. The following sections describe the physical and technical details of all the components used in instruments for projection-type imaging, such as the x-ray source, imaging optics, detectors, and image processing.

3.2 X-ray sources

3.2.1 Synchrotrons

As illustrated in figure 3.2, synchrotron radiation is an electromagnetic wave emitted by electrons and/or positrons traveling in an orbit, typically the orbit of a synchrotron or a storage ring, which were originally developed as tools for experimental fundamental physics [1–8]. In such an accelerator, the electrons are accelerated to relativistic speed and have very high energy. For example, at the European Synchrotron Radiation Facility (ESRF) the electron energy $\mathcal{E}_e$ is 6 GeV, and this value is as high as 8 GeV at SPring-8 in Japan. Based on the equation

$$\mathcal{E}_e = \frac{mc^2}{\sqrt{1 - v^2/c^2}}, \tag{3.1}$$

the speed of an electron can come extremely close to the speed of light, namely $(c - v)/c = 2 \times 10^{-9}$ when $\mathcal{E}_e = 8$ GeV. In science fiction, miracles may happen when approaching the speed of light, but in synchrotrons, such high speeds have become a reality. Researchers are taking advantage of this to solve many physical, chemical, biological, and other scientific problems closely related to daily life!

The radiation produced by bending magnets has a continuous spectrum. The critical frequency ω_c is determined by the electron energy in the storage ring and the magnetic strength:

$$\hbar\omega[\text{keV}] = 0.665\mathscr{E}_e^2[\text{GeV}]B[\text{T}]. \tag{3.2}$$

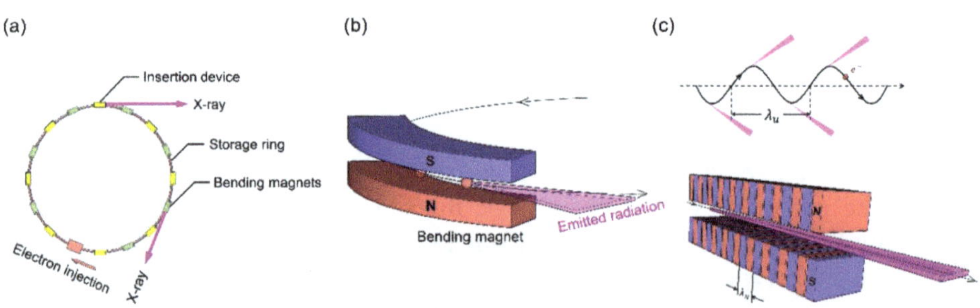

Figure 3.2. High-energy electrons traveling in a storage ring (a) are channeled by bending magnets (b) or periodic magnets in insertion devices (c) and emit electromagnetic radiation. In (c), λ_u is the period of the magnet pairs. Further details are described in [1–8].

In general, insertion devices are classified into two types: undulator and wiggler, depending on their K-factor:

$$K = \frac{eB\lambda_u}{2\pi m_e v}, \tag{3.3}$$

where q is the charge of an electron, B is the magnetic field of the ID, λ_u is the period of the magnetic pairs, and m_e and v are the mass and the speed of the electron, respectively. Undulators have a K-factor much smaller than one, but for wigglers, $K \gg 1$.

When $K \ll 1$, the oscillation amplitude of the electron motion is small, and the radiations generated by different oscillations may interfere with each other. Therefore, undulator x-rays may present some coherent properties.

When $K \gg 1$, the oscillation amplitude is too large to allow interference to happen. The trajectory of an electron is like a series of circular arcs. The emitted radiation is linearly enhanced by adding more magnet pairs. The spectrum is similar to that produced by bending magnets.

A double-crystal monochromator is frequently utilized to get monochromatic synchrotron radiation. As shown in figure 3.3, it is generally horizontally placed and consists of twin crystals. In most cases, the crystals are silicon, high-quality crystals of which are relatively easy to obtain. The crystal reflection plane is typically silicon (111), (220), or (311). In practice, the selection of a reflection plane is mainly based on the reflection angle. For example, when the energy of the incident beam is 10 keV and its wavelength is 1.24 Å, the reflection angles are 11.4°, 18.8°, and 22.2° for silicon (111), (220), and (311), respectively. Sometimes, the double-crystal monochromator is rotated to get x-rays with tunable wavelength.

The bandwidth of the monochromatic radiation can be estimated as follows:

$$\frac{\Delta\lambda}{\lambda} = \frac{\Delta E}{E} = \cot(\theta)\Delta\theta, \tag{3.4}$$

where θ is the reflection angle and $\Delta\theta$ is the angular divergence of the monochromatic radiation. $\Delta\theta$ is given by

$$\Delta\theta = [(\Delta\psi)^2 + (\omega_D)^2]^{\frac{1}{2}}, \tag{3.5}$$

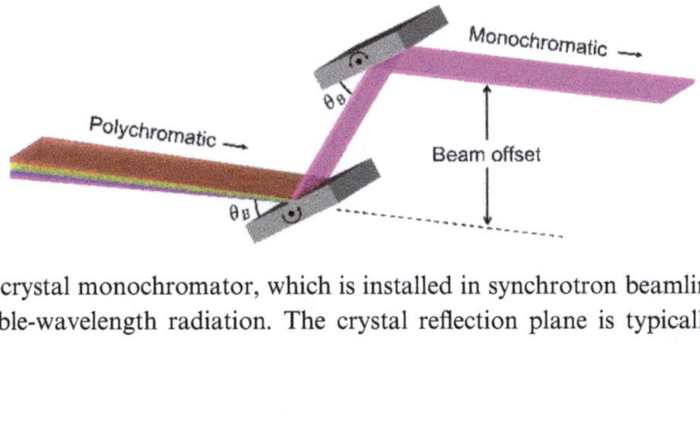

Figure 3.3. A double-crystal monochromator, which is installed in synchrotron beamlines as almost standard optics, provides tunable-wavelength radiation. The crystal reflection plane is typically silicon (111), (220), or (311).

where $\Delta\psi$ is the beam divergence from the x-ray source (which is mainly controlled by the emittance of the x-ray source, the source-to-monochromator distance, and the width of the incident slit in front of the monochromator) and ω_D is the Darwin width, which depends on the x-ray wavelength and the reflection plane and can be calculated via the dynamic x-ray diffraction theory. For higher crystal indexes, ω_D is smaller. Based on equation (3.5), we can see that both $\Delta\theta$ and the bandwidth of the monochromatic radiation become smaller when the indexes of the reflection plane are larger. Sometimes this impact is considered in the selection of the crystal reflection plane. The bandwidth $\Delta E/E$ is typically 10^{-4} to 10^{-3}. It is worth noting that it is sufficient to have the first crystal to realize most functions of a monochromator. The introduction of the second crystal is mainly to guide the monochromatic radiation so that it propagates in the original horizontal direction.

For the purpose of projection-type XRF imaging, the instrument layout needs to be specially designed. When the target elements have a low concentration and are diluted in a matrix, the x-rays scattered by the matrix are much stronger than the fluorescent x-rays produced by the target elements, resulting in a high background in the resulting XRF images. Since the angular distribution of the scattering intensity has a sharp minimum in the polarization direction of the primary x-rays, it is preferable to align the XRF detector in the direction of polarization. This layout has become nearly a standard in synchrotron XRF analysis, as shown in figure 3.4(a).

Synchrotron radiation is typically horizontally wide but vertically narrow and is polarized in the horizontal direction. To fully illuminate a large area of the sample, a natural design is to align both the sample and the camera in the horizontal direction, as shown in figure 3.4(b). In this case, the camera is in the direction perpendicular to the polarization, making it receive the most scattering x-rays and resulting in the strongest background. To reduce the scattering background and to improve the signal-to-background ratio in XRF images, a better layout is to align both the sample and the camera in the vertical direction. In this case, only a thin strip (typically 1–2 mm wide) of the sample can be illuminated, as shown in figure 3.4(c). One may consider expanding the beam in the vertical direction using an asymmetrically cut crystal, as reported in reference [9].

3.2.2 X-ray tubes

An x-ray tube is a vacuum tube that can produce x-rays based on the impact of high-energy electrons on a metal target [7, 10, 11]. As shown schematically in figure 3.5, it consists of a cathode, which is an electron source, and an anode, which is a metal target. Both are in vacuum, which in many cases is in a sealed glass tube. The typical cathode uses a hot electron emission source, which is a filament that is heated to a high temperature by electric current. In the case of a conventional tungsten filament, the temperature can be about 2000 °C. The electrons are accelerated by a high voltage applied between the cathode and the anode. X-rays are emitted when the high-speed electrons strike the metal target at the anode and are then collected through an x-ray window.

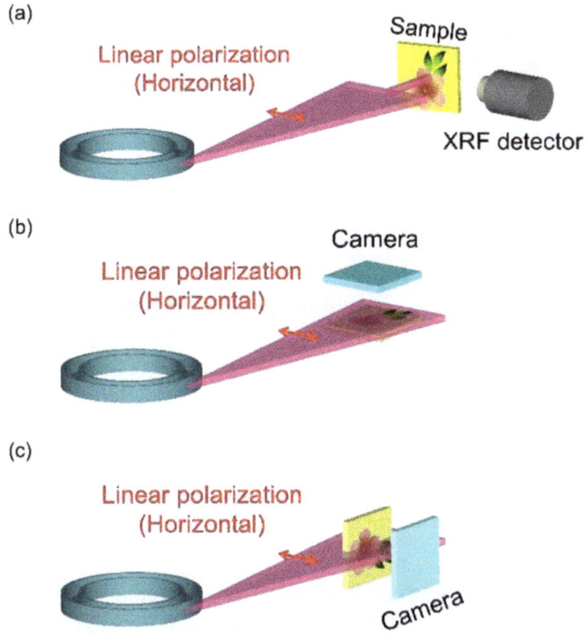

Figure 3.4. Polarization and geometry in synchrotron experiments [9]. (a) In a typical synchrotron XRF measurement, the detector is aligned with the direction of polarization to reduce scattering background. (b) In synchrotron projection-type XRF imaging, the sample and the camera can be horizontally aligned to use the wide synchrotron beam, but in this case the scattering background is the strongest. (c) To reduce scattering background, it is recommended to place the sample and the camera vertically and expand the synchrotron beam in the vertical direction.

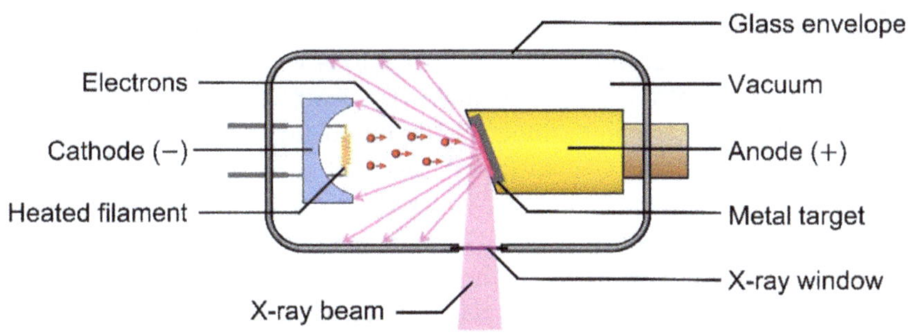

Figure 3.5. A schematic illustration of an x-ray tube. Electrons are emitted by a high-temperature filament and then accelerated by an electrical potential, typically tens of kilovolts, between the anode and the cathode. X-rays are generated when the electrons collide with a metal target. Further details are given in [7, 10, 11].

Frequently, the area of the metal target heated by electrons has a rectangular shape whose length is denoted by L and whose width is denoted by H. The shape of the x-ray beam depends on the geometry of the x-ray window. There are two basic configurations, as shown in figure 3.6. If the x-ray window is aligned with the longer

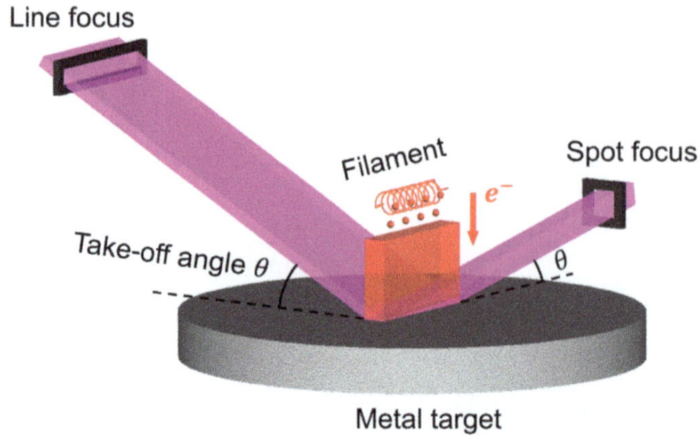

Figure 3.6. The electron-heated region on the metal target is typically in a shape of long and narrow rectangle. When x-rays are taken from different directions, the x-ray source appears to have different shapes. Line focus is recommended for experiments with a crystal monochromator and/or mirrors that include projection-type XRF imaging. Further details are described in [7, 10, 11].

side, the shape of the x-ray beam is $L \sin\theta \times H$, where θ is the take-off angle. This geometry is called 'spot focus' and it is frequently adopted for single-crystal diffraction. On the other hand, if the x-ray window is aligned with the shorter side, the x-ray beam takes on the shape of $L \times H \sin\theta$. This geometry is called 'line focus.' For the purpose of projection-type XRF imaging, it is better to use the line focus geometry to grazingly illuminate a wide area on the sample. In addition, there is another geometry named 'microfocus,' where the x-ray window is at the back of, or very close to the heated area on the metal target. In this case, the emitted x-ray beam covers a wide range of angles. This microfocus geometry is frequently adopted by radiography (x-ray transmission imaging) or low-power x-ray tubes. When necessary, it can also be applied to projection-type XRF imaging, whereas collimation or monochromatization of the x-ray beam is not easy.

When high-speed electrons impinge on the metal target, two types of x-rays are generated, corresponding to two different physical processes. A small number of electrons are decelerated by inelastic x-ray scattering, giving rise to a continuous spectrum, known as bremsstrahlung (a German word meaning 'braking radiation'). Its upper energy limit corresponds to the tube voltage. In other interactions, if the incident electron has an energy higher than the binding energy of an electron in an inner orbit, the inner electron may be removed, resulting in cascading electron transitions and giving rise to fluorescence x-rays, the energy of which is characteristic of the metal target. For example, figure 3.7 plots the XRF spectra of a molybdenum target. The spectra were simulated using the package 'SPekPy' [12]. As the K-absorption edge of molybdenum is 20 keV when the tube voltage is 20 kV, the characteristic molybdenum Kα (17.478 keV) and Kβ (19.607 keV) lines are nearly invisible, and there is only continuous bremsstrahlung, the highest energy of which is 20 kV. When the tube energy is higher than 20 kV, characteristic molybdenum lines become visible.

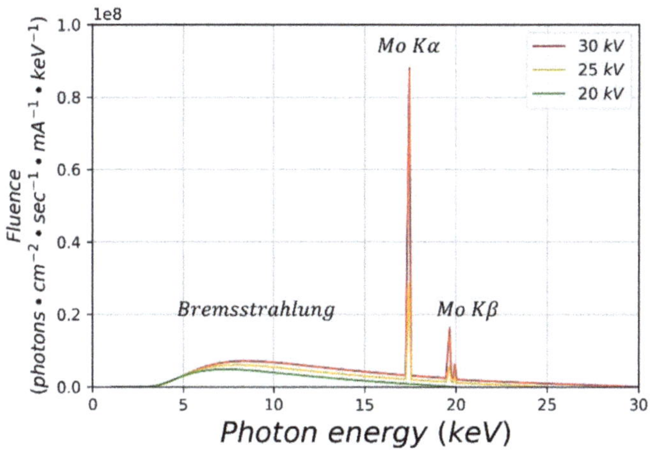

Figure 3.7. Calculated x-ray tube spectra of a molybdenum target attenuated by 20 cm of air (simulated by the software, 'SPekPy' [12]). When the tube voltage is higher than the molybdenum's K-absorption edge (around 20.0 keV), characteristic molybdenum Kα and Kβ x-rays are generated. In the simulation, the take-off angle is set to 12 °.

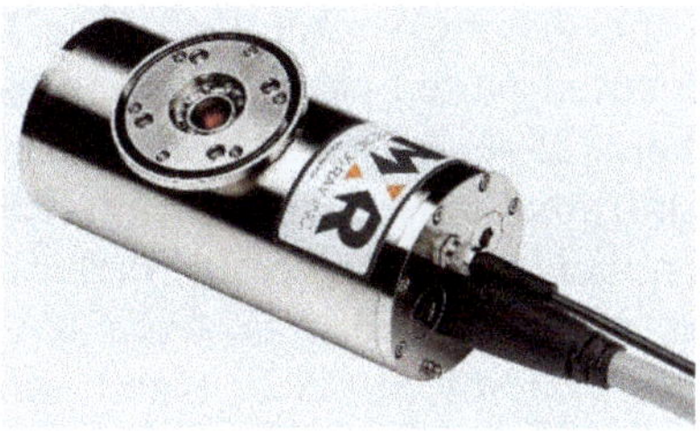

Figure 3.8. A mini-focus packaged x-ray tube from MICRO X-RAY Inc. [23]. Reprinted with permission from www.microxray.com.

In an x-ray tube, the available x-ray intensity is roughly limited by the fact that the metal target can be melted by too strong an electron beam. To obtain more x-ray photons, it is helpful to use a rotating anode x-ray source [13–17] or a liquid metal jet anode x-ray source [18–20]. On the other hand, in the last decade, miniature x-ray tubes suitable for portable XRF spectrometry have gradually become popular [21]. Most of them adopt the microfocus geometry and are low-powered, which makes water cooling unnecessary. These systems would be useful not only for ordinary energy-dispersive XRF analysis but also for projection-type XRF imaging, which gives maps of each chemical elements contained in the sample. Such miniature x-ray tubes have become commercially available [22, 23] and offer new possibilities for XRF imaging in the near future. Figure 3.8 shows one of them.

Continuous bremsstrahlung may give rise to a scattering background in XRF analysis. One may reduce its intensity by inserting a filter into the primary x-ray beam. For example, for the tube spectra of the molybdenum target in figure 3.7, a thin zirconium filter is appropriate. As the K-absorption edge of zirconium is 17.998 keV, it may block the continuous spectrum up to 17.998 keV in figure 3.7 and also reduce the intensity of molybdenum Kβ. When the tube voltage is much higher than 20 keV, one may also consider using a reflection mirror to remove the high-energy continuous spectrum. On the other hand, if the primary x-ray beam is well collimated, it is also possible to use a monochromator. The monochromator can be a single crystal of silicon or lithium fluoride, or curved graphite. The passbands of lithium fluoride and curved graphite are generally wider than that of silicon.

3.2.3 Laser-driven x-ray sources

The x-rays produced by conventional x-ray tubes are time-continuous electromagnetic waves, but the use of short-pulse x-rays is desirable for studies that follow changes in a sample on very short timescales. Pulsed lasers with nanosecond, picosecond, and femtosecond pulse widths are widely used in the wavelength range from infrared to visible and ultraviolet light. If such pulsed lasers become available in the x-ray wavelength range, it will be possible to add a high degree of time resolution to existing analytical methods based on x-rays. Various techniques have been developed and applied to generate pulsed x-rays using high-performance infrared pulsed lasers as the driving source [24]. The laser-plasma x-ray source irradiates a target with a focused pulsed laser beam and uses the radiation produced by the plasma generated there [25–27]. Inverse Compton scattering x-ray sources use x-rays emitted when an infrared pulsed laser beam collides with a high-energy electron beam [28, 29]. As noted in the 2023 Nobel Prize study on attosecond science [30], higher harmonic generation (HHG) by an inert gas irradiated by an infrared femtosecond laser produces extreme ultraviolet and soft x-ray pulses with pulse widths even shorter than those of the original laser [31, 32]. Ultrashort pulsed x-rays in the keV range can now be obtained and used for time-resolved spectroscopic studies [33]. On the other hand, the world's first 8 keV x-ray free-electron laser (XFEL) was successfully launched in 2009 at the SLAC National Accelerator Laboratory in the USA [34].

All of the above technologies continue to advance, and it is expected that pulsed x-ray sources will soon be available in an even more user-friendly form. In addition to time-resolved x-ray fluorescence spectroscopy, there are great expectations for the development of x-ray color imaging that can be used to see the spatial distributions of chemical elements in a sample. For such applications, projection x-ray fluorescence imaging will be the only option, and scanning techniques will not be suitable.

3.3 Imaging optics

In order to perform projection-type XRF imaging, x-ray optical components are required to project images from the sample onto the camera. However, unlike visible light, there are no converging lenses available for x-rays. The sophisticated optical systems designed for visible light cannot be applied to x-ray imaging. Generally speaking, there are two types of optical components that are commonly used for x-ray imaging: collimators and reflection mirrors.

A collimator transmits x-rays propagating in specific directions and blocks x-rays spreading in other directions. Typically, collimators have a trade-off between transmitted intensity and spatial resolution. However, they are usually robust and reliable in practical experiments. When designing collimators, careful consideration must be given to the selection of shielding materials and their thickness.

The intensity of x-rays transmitted through a material is given by the equation:

$$I = I_0 \exp -\mu t, \tag{3.6}$$

where t is the thickness of the material and μ is its linear absorption coefficient. At a given x-ray energy, μ increases for high-Z and high-density materials. For a given material, as the x-ray energy increases, the corresponding μ decreases. Figure 3.9 shows the thicknesses of different materials required to shield x-rays down to $I/I_0 = 10^{-4}$. The step-like abrupt changes in the curves are due to the absorption edges of the materials. Clearly, the efficiency of absorption is higher for high-density and high-Z elements such as tungsten, platinum, and lead. Therefore, it is recommended to use these materials to fabricate or coat x-ray collimators.

Reflection mirrors are widely used in x-ray imaging. This is because x-rays can be totally reflected at grazing incidence by a smooth surface. In contrast to visible light, the refractive index of x-rays in materials is smaller than unity [35]. It should be clarified that this does not imply that the speed of light in the material is greater than that in vacuum, as the velocity of an electromagnetic wave in a material is c/n. In fact, c/n represents the phase velocity, whereas the group velocity, given by $\mathrm{d}\omega/\mathrm{d}k$, is always less than c.

Therefore, when an x-ray beam is incident from vacuum or air into a denser material, its refractive behavior differs from that of visible light, as illustrated in

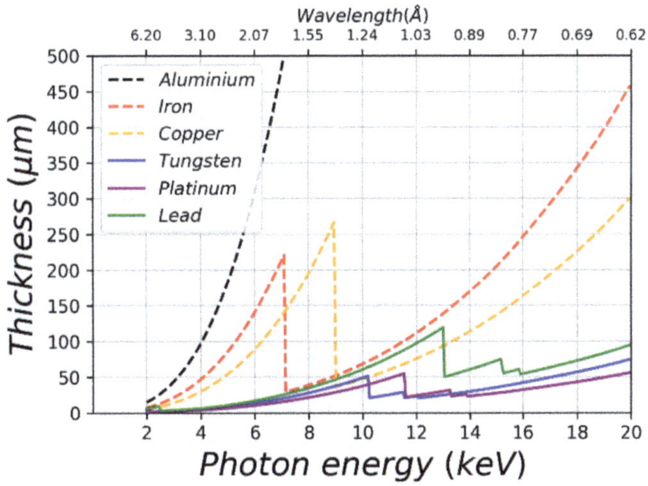

Figure 3.9. The thicknesses of different materials required to block x-rays down to 10^{-4}.

figures 3.10(a) and (b). Specially, when the grazing angle α is smaller than a critical angle α_c, the incident x-ray beam can be totally reflected, as shown in figure 3.10(c). The critical angle α_c is determined by the x-ray wavelength and the material's electron density:

$$\alpha_c \approx \lambda\sqrt{\frac{\rho r_0}{\pi}}, \tag{3.7}$$

where λ is the wavelength, ρ is the number density of electrons, and $r_0 = 2.81794 \times 10^{-15}$ m is the classical electron radius, also known as the Thomson scattering length or the Lorentz radius.

Figure 3.11 displays the critical angles of x-rays incident on the surfaces of different materials. As the x-ray wavelength decreases, the critical angle also decreases. Note that the critical angle is given in milliradians, which is a very small value. As a result, to focus x-rays, the curvature of the x-ray mirror must be small, and the corresponding focal length must be large. It is also noteworthy that denser

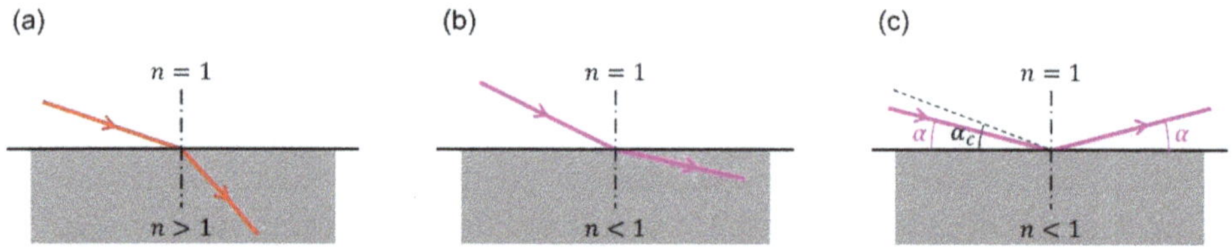

Figure 3.10. The reflection and refraction of electromagnetic waves at an interface [35]. (a) The refraction of visible light. The refractive index of denser materials is considerably greater than unity. (b) The refraction of x-rays. The refractive index of denser materials is slightly less than unity. (c) Total reflection happens when x-rays are incident from vacuum or air onto the surface of denser materials and the glancing angle α is smaller than the critical angle α_c.

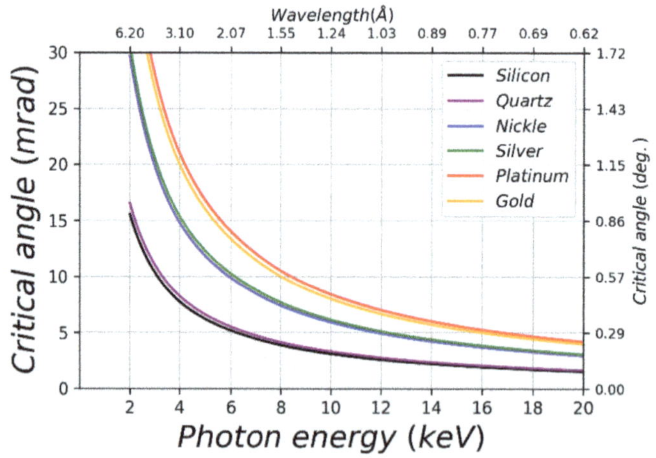

Figure 3.11. The calculated critical angles of x-rays of different wavelengths at the surfaces of different materials.

materials have higher critical angles. Therefore, while most x-ray mirrors are made of silicon or quartz, it is still advisable to coat the mirror surfaces with platinum or gold. To achieve optimal focusing performance, the surface shape of x-ray mirrors must be meticulously designed and fabricated, while the surface roughness must be rigorously controlled.

There are also some other types of x-ray optical components, including single-crystal or multilayer reflection mirrors, Fresnel zone plates [36], and multilayer Laue lenses [37]. However, these components only support wavelength-selective imaging and are unsuitable for achromatic projection-type XRF imaging intended to simultaneously measure the distributions of different elements. Therefore, we will not discuss them in this book.

3.3.1 The pinhole optic

The use of a pinhole to project images has a long history. Pinhole imaging is a simple physical phenomenon. Its descriptions can be found as early as the book 'Mozi' from the 5th century in China and also in the writings of Aristotle (384–322 BC). During the Middle Ages and the Renaissance period, pinhole imaging was often used as an aid in painting. In 1856, in the context of modern photography, David Brewster published the physical explanation of stereoscopy, including the description of the idea as 'a camera without lenses, and with only a pin-hole' [38]. It was also used in the form of the pinhole camera in early camera obscuras. In modern times, with the development of optical lenses, pinhole imaging is rarely used in the domain of visible light. However, even today, as one of the simplest and cheapest imaging optics, pinhole imaging is still widely used in x-ray imaging.

Pinhole imaging follows very simple ray optics, as shown in figure 3.12. Theoretically, diffraction occurs when the pinhole size approaches the wavelength of light. In this case, ray optics is no longer applicable. However, since the wavelength of x-rays is on the angstrom scale, we do not need to worry about this in the domain of x-rays.

As shown in figure 3.12, when the distance from the object to the pinhole is d_o and the distance from the image to the pinhole is d_i, the imaging magnification is

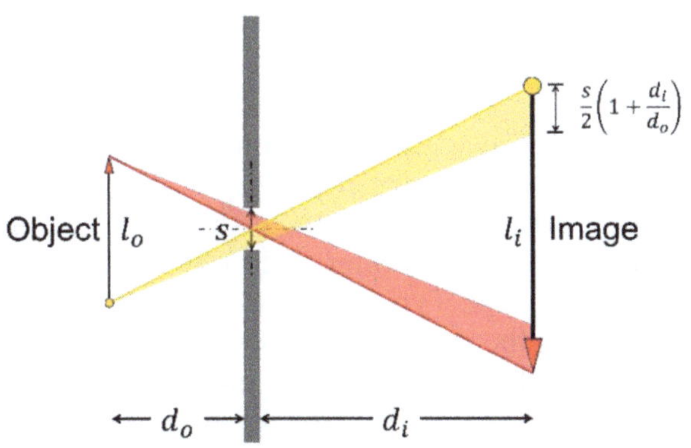

Figure 3.12. A schematic illustration of pinhole optics.

$M = d_i/d_o$. When the distance between two points on the object is less than a certain distance, these two points cannot be distinguished in the image. This distance is the spatial resolution Δ of the imaging method. Assuming the thickness of the pinhole plate is negligible compared with the pinhole size, R is given by

$$\Delta = \left(1 + \frac{1}{M}\right)s. \tag{3.8}$$

Clearly, the spatial resolution of pinhole imaging R is dominated by the pinhole size s. Furthermore, R approaches s as the imaging magnification M increases.

In practice, the required thickness of the pinhole plate is the main limiting factor for achieving high resolution in pinhole imaging. As plotted in figure 3.9, the pinhole plate needs to have a certain thickness to block x-rays down to 10^{-4}. Tungsten is probably the most suitable material for making the pinhole plate, considering its mechanical properties and x-ray shielding capability. To shield fluorescence x-rays below 20 keV, a thickness of 25–50 μm is required. The pinhole size should not be significantly smaller than the thickness to ensure an adequate pinhole opening angle. When the pinhole size approaches the thickness, the optics becomes more complicated than that shown in figure 3.12, and the spatial resolution cannot be accurately predicted using analytical expressions. It is better to run numerical simulations or test standard resolution targets experimentally, although the spatial resolution is generally not far from the pinhole size. Typically, the spatial resolution of pinhole imaging ranges from 50 to 100 μm. To the best of the authors' knowledge, the best resolution [39] achieved so far is 15 μm, using a pinhole 10 μm in diameter on a 50 μm thick tungsten foil and an imaging magnification of 8.5 times for chromium patterns.

If a pinhole larger than 200 μm is needed, it is easy to manufacture it manually on a tungsten foil using a pin. However, if a very small pinhole, such as one below 50 μm, is needed, manufacturing equipment is still required. For example, the authors have used a 266 nm ultraviolet laser. During the manufacturing process, the tungsten foil was precisely adjusted to the laser focus and the number of laser pulses was carefully controlled to make the pinhole size as small as possible. The pinhole size can be inspected using an optical microscope and x-ray transmission imaging. This process of manufacture and inspection can be completed in as little as 10 min. It is much easier and cheaper than manufacturing any other x-ray optical component.

Apart from being easily obtainable and cost-efficient, there are other advantages to using a pinhole in XRF imaging. In pinhole imaging, there is no optical focusing, and therefore the object distance, image distance, and imaging magnification can be continuously adjusted as desired. Aligning the instruments is quite easy. In general, it is only necessary to ensure that the center of the object, the pinhole, and the center of the detector sensor are coaxial. Typically, in an image projected by a pinhole, the central area is brighter than the surrounding areas. This is due to the different solid angles caused by different distances from the pinhole. However, this solid angle difference can easily be corrected. No additional image processing is required. Because of these advantages, pinholes have been widely used in both pioneering works on XRF imaging and in many current experiments. Figures 2.5, 2.10, 2.14, 2.15, 2.16, 2.20, 2.21, 2.26, and 2.31 in chapter 2 were obtained using pinholes.

The most obvious limitation of pinhole imaging is its imaging efficiency. As mentioned in the above text, collimator-based optics, which allows x-rays propagating in specific directions to pass while blocking x-rays spreading in other directions, involves a trade-off between the transmitted intensity and the spatial resolution. Pinhole imaging is a typical example of this. To obtain high-resolution images, a small pinhole size is needed, resulting in a small solid angle over the object and a low intensity of transmitted x-rays. As a result, a relatively long time is needed to accumulate a high-quality XRF image with sufficient signal. This limitation has prompted researchers to consider other collimator-based x-ray imaging optics.

3.3.2 The coded aperture

If one is not satisfied with the solid angle of a single pinhole over the object or the transmitted x-ray intensity, a natural idea is to increase the number of pinholes so that the transmitted intensity is multiplied. As shown in figure 3.13, each pinhole on the plate can individually project an independent image onto the detector sensor, and these images overlap. The arrangement of pinholes on the plate is preferably specially designed, making the plate a so-called coded aperture. The projected overlapping image needs to be processed by algorithms to reconstruct the original single image. The idea of using multiple pinholes was first developed in the field of astronomy for x-ray and gamma-ray imaging over half a century ago [40, 41]. It was then further developed into the more complex coded aperture [42] and eventually spread to other fields [43]. In the last 15–20 years, it has also been applied to XRF imaging [44–48].

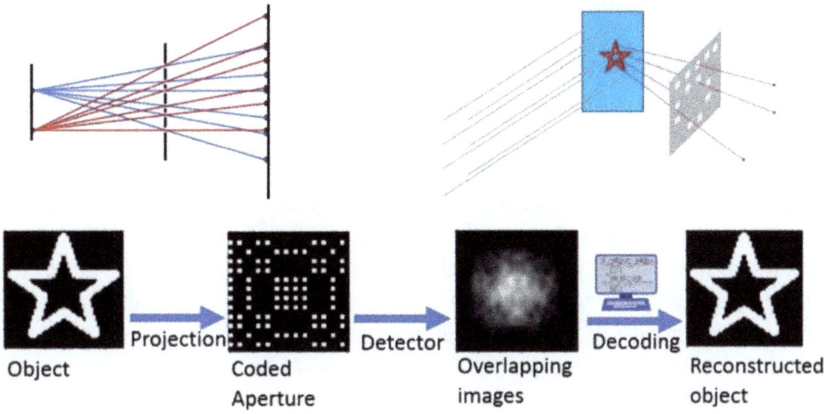

Figure 3.13. A schematic illustration of coded-aperture XRF imaging. The coded aperture can be considered to consist of multiple pinholes. Each pinhole can individually project an XRF image, and all the images overlap. Image-processing algorithms are required to reconstruct the original single image. Adapted from figures 1 and 2 in reference [45] with permission from the Royal Society of Chemistry.

In coded-aperture XRF imaging, the development of a reliable image reconstruction algorithm poses the greatest challenge [45, 46]. Conventionally, image reconstruction is a simple 'decoding' process achieved by convolution with an antimask. The construction of the antimask is associated with the design of the coded mask. This decoding method is mathematically well defined and widely applied in the field of astronomy. However, in astronomy, the object is assumed to be infinitely far away, and the incident light is parallel. In contrast, in XRF imaging, the object is placed in the near field of the coded aperture, resulting in a magnification of the projected image. Therefore, scaling must be performed before convolution with the antimask. Furthermore, the near-field effect also introduces the thickness of the coded aperture as a disturbing factor. The decoding method can still be perfectly applied if the coded aperture has no thickness. However, its performance in reconstruction greatly degrades as the thickness of the coded aperture increases, such as to 100 μm.

Newly developed image reconstruction algorithms for near-field XRF imaging include iterative optimization algorithms and machine learning algorithms. The basic idea of iterative optimization is trial and error. First, an arbitrary object is assumed, and its corresponding image projected via the coded aperture is simulated. The simulated image is then compared with the actual measured image, and based on the differences, the assumed object is optimized through negative feedback. After 100 to 200 iterations, the original object can be successfully reconstructed. The machine learning algorithm requires a training data set that includes various objects and their corresponding simulated projected images. If the projection function of the coded aperture in XRF imaging is fully premeasured and characterized, iterative optimization algorithms and machine learning algorithms may outperform the antimask method in terms of reconstruction performance.

To demonstrate, figure 3.14 presents an example of coded-aperture XRF imaging. In figure 3.14(a), the design of the coded aperture is shown. It consists of twelve 90 μm (dia) pinholes manufactured on a 25 μm thick platinum foil. The sample comprised an outermost square of cobalt, an inscribed circle of nickel, a large upright triangle of copper, and a small inverted triangle of iron in the innermost region. The experiment was carried out at BAMline at the synchrotron BESSY II, Germany, utilizing 15 keV monochromatic primary x-rays. A p–n charge-coupled device (CCD) detector was employed. The recorded images exhibit overlapping patterns, but the original images of the different elements can be reconstructed using the abovementioned methods: the antimask method, the iterative optimization algorithm, and the machine learning algorithm.

Among the major advantages of coded-aperture XRF imaging are its large solid angle and relatively good efficiency in collecting XRF signals, while difficulties may remain in image-processing algorithms for inhomogeneous and irregular natural samples.

3.3.3 The collimator plate

A pair of microchannel plates is often used to detect particles such as electrons and ions [49]. Such a device is closely related to the electron multiplier and provides spatial resolution by having many separate channels. In the microchannel plate

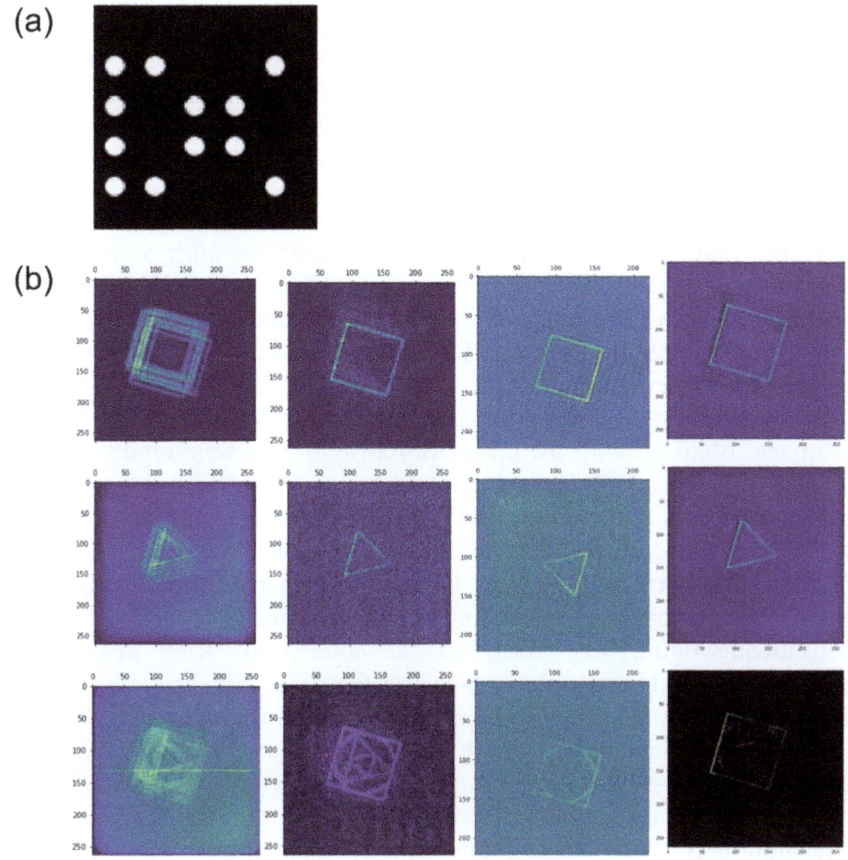

Figure 3.14. An experimental demonstration of coded-aperture XRF imaging. (a) The design of a coded aperture consisting of twelve pinholes 90 μm in diameter. (b) Detected and reconstructed images. The first, second, and third rows correspond to cobalt Kα, copper Kα, and the combination of all energy regions, respectively. The first column shows the images recorded without any image processing. The second, third, and fourth columns display the reconstructions obtained using the antimask method, the iterative optimization algorithm, and the machine learning algorithm, respectively. Adapted from figures 5 and 12 in reference [46] with permission from the Royal Society of Chemistry.

detector, each microchannel plate is electrically connected. Amplified electrons are collected at the anode, resulting in the signal event. On the other hand, a single microchannel plate can be used as a collimator plate for x-rays. Since it is not used for charged particles, no wiring is required. Therefore, even the raw material of the microchannel plate can also be used as an x-ray collimator plate. When higher energy x-rays are used, a gold or rhodium coating is desirable to prevent transmission through the wall of each channel.

In short, a collimator plate is a component that densely assembles numerous parallel straight microchannels into a plate. It allows only x-rays propagating along its axis to pass through, while blocking x-rays propagating in other directions. Typically, a microcapillary has a diameter of 6 μm and the plate thickness is 1 mm. The corresponding maximum angular acceptance is 6 mrad. By the late 1980s, the feasibility of using a single microchannel plate as a collimator plate in x-ray optics was well recognized [50]. When used as a collimator for multilayer mirrors and/or analyzing crystals, the system can provide a 2D monochromatic x-ray image [50–52]. Therefore, in principle, the system can be applied to wavelength-dispersive XRF spectroscopy, although deformation occurs depending on the diffraction angle of the spectrometer.

In 1995, T. Wroblewski reported a pioneering work on the imaging of x-rays from a sample, rather than from some x-ray optical components such as mirrors, by the combined use of a single microchannel plate and a CCD camera [53]. This work used x-rays diffracted by a polycrystalline sample. He also pointed out its potential usefulness in x-ray absorption spectroscopy imaging [54]. At that time, one of the authors, K Sakurai, met T Wroblewski in Hamburg and discussed the possibility of applying it to XRF imaging. The first experiment in Sakurai's laboratory was attempted in 1997, and the first images obtained with the laboratory x-ray source were reported at the 1998 international conference [55]. Instead of a commercially available microchannel plate product, a much cheaper collimator plate, which is a raw material of microchannel plates, was used. To prevent x-ray transmission, some metals such as gold and/or rodium are coated on the entire surface of the plate, including the channel parts. Then, many XRF images were obtained by directly installing a collimator plate into a 2D detector [56, 57].

The following figure 3.15 illustrates the concept of XRF imaging using a collimator plate [57]. The sample surface and the detector sensor are placed nearly parallel, while the sample needs to be slightly tilted to receive a thin primary x-ray beam. A collimator plate is inserted into the gap between the sample surface and the detector sensor. Since it only passes fluorescence x-rays parallel to its axis, an equal-sized XRF image is projected onto the detector sensor by the sample surface. The spatial resolution Δ of the image is approximately given by

$$\Delta = d \times \frac{r}{t}, \tag{3.9}$$

where r is the capillary diameter, t is the collimator plate thickness, and d is the distance from the sample surface to the detector sensor. According to this equation, when the diameter-to-length ratio r/t of the capillary is a constant, d should be minimized to achieve the highest possible spatial resolution. Typically, when r is 6 μm and t is 1 mm, it is better to reduce d to less than 3 mm to obtain a spatial resolution better than 20 μm. To achieve this, the collimator plate should be installed inside the detector housing, directly on the surface of the sensor. The front end of the detector should be in close proximity to the sample surface. In some cases, it may be necessary to modify the detector housing to position the sensor near the front end. This strict requirement on the sample-to-detector distance necessitates that the sample surface must be flat. This may limit the use of this method with uneven or reactive samples.

In the case where the diameter-to-length ratio of a capillary on the collimator plate remains unchanged, the solid angle of the capillary over the sample surface is constant, regardless of the distance between them. Therefore, changing the distance between the collimator plate and the sample does not affect the transmitted x-ray intensity or the imaging efficiency. However, as the distance increases, the spatial resolution degrades.

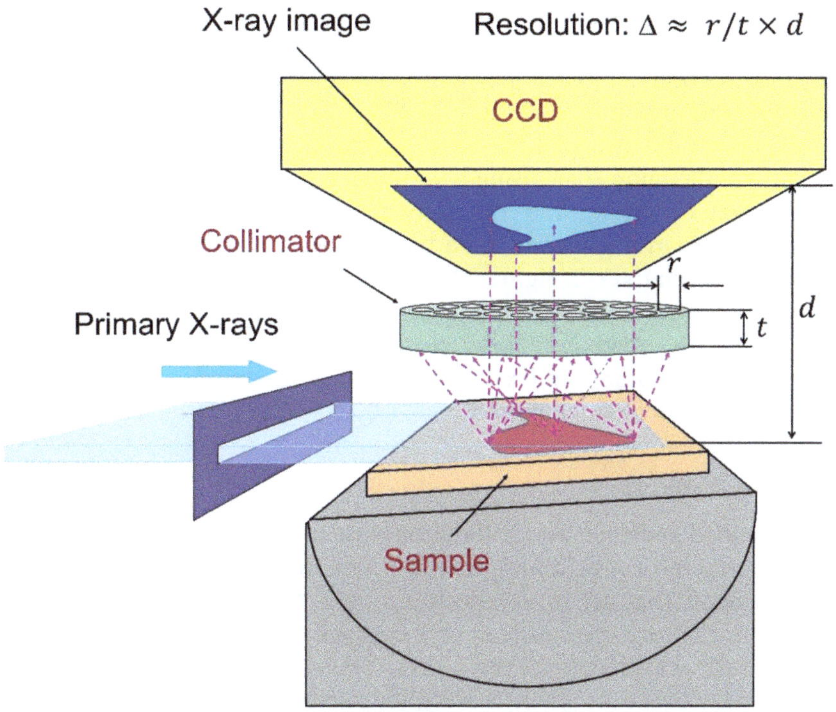

Figure 3.15. A schematic illustration of XRF imaging using a collimator plate. The gap width d between the sample surface and the detector sensor should be as small as possible [55], copyright IOP Publishing Ltd. All rights reserved.

Consequently, at a large distance, the projected image is severely smeared, making it impossible to observe any details or features. This leads to the failure of XRF imaging.

Another factor that affects the transmitted x-ray intensity is the opening ratio of the collimator plate. This refers to the ratio of the total opening area of all the capillaries to the area of the collimator plate. The opening ratio is determined by the cross-sectional shape of the capillary and the thickness of the walls between the capillaries. Its magnitude has a linear impact on the transmitted x-ray intensity. Typically, the opening ratio falls between 0.5 and 0.6.

The thickness of the walls between the capillaries determines the upper limit of x-ray energy that can be used with the collimator plate. When the energy of the x-rays exceeds this limit, they bleed into adjacent capillaries, resulting in smearing of the projected XRF image. It is worth noting that when using an energy-dispersive x-ray detector, high-energy x-ray images may smear, whereas low-energy x-ray images remain unaffected. The material typically used for a collimator plate is lead glass or silica glass. To enhance its ability to block x-rays, it is recommended to coat a gold layer on the inner wall of the capillary.

It should be noted that total reflection occurs when the x-ray grazing angle on the capillary inner surface is smaller than the critical angle. The values of the critical angle for different materials are plotted in figure 3.11. Since the acceptance angle of the capillary is 6 mrad, which is greater than or close to the critical angle, the proportion of totally reflected x-rays is negligible, and we only need to consider the directly transmitted x-rays. This is often confused with polycapillary optics, which is introduced in section 3.3.6. In polycapillary optics, totally reflected x-rays are dominant. We will continue to discuss the difference between them in section 3.3.6.

When using a collimator plate in XRF imaging, adjusting the instrument can be slightly more challenging than using a pinhole, but it is still relatively straightforward. At the time of writing, the collimator plate has been the main technique used by the authors' research team for more than 20 years. Its application spans a wide range of fields. In chapter 2, figures 2.2, 2.7–2.9, 2.11, 2.13, 2.23–2.25, 2.27–2.29 were acquired using a collimator plate. In addition, the authors have applied the collimator plate to other advanced x-ray imaging techniques, such as x-ray diffraction (XRD) imaging and x-ray absorption fine structure (XAFS) imaging. These will be introduced in chapter 4.

3.3.4 The Wolter mirror

The pinhole, coded aperture, and collimator plate mentioned in the previous sections all follow the principles of collimation optics. The following sections introduce x-ray optical components that use total reflection optics. One of the major advantages of using total reflection of x-rays instead of Bragg reflection is that the focal point does not depend on the wavelength of the x-rays. This is particularly important in x-ray spectroscopy. Our first example is the Wolter mirror. It was designed by H. Wolter in 1952 for x-ray telescopes [58]. There are three types of Wolter mirrors: type I, type II, and type III. They differ significantly in the arrangement of their two reflecting mirrors. Each type has its own advantages and disadvantages when used in telescopes. Type I is suitable for x-ray microscopes and can be used for both focusing and imaging optics.

The optics of a Type I Wolter mirror is shown schematically in figure 3.16. It consists of two reflecting mirrors: a hyperboloid mirror (shown in blue) and an ellipsoid mirror (shown in green). The left focus and right focus of the hyperboloid

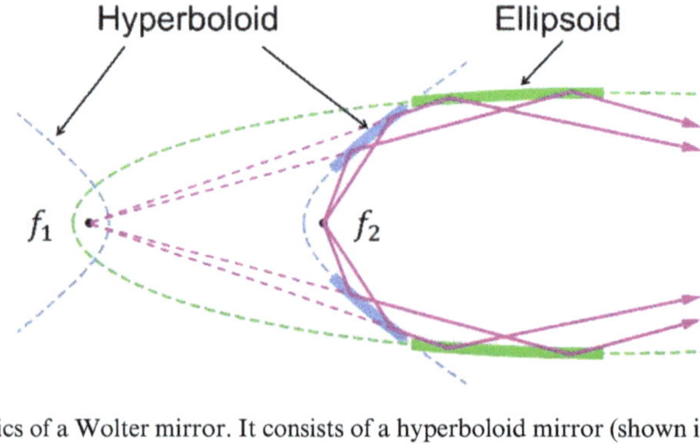

Figure 3.16. The optics of a Wolter mirror. It consists of a hyperboloid mirror (shown in blue) and an ellipsoid mirror (shown in green). f_1 and f_2 represent the left focus and the right focus of the hyperboloid mirror, respectively. f_1 is also the left focus of the ellipsoid mirror. Fluorescence x-rays (shown in magenta) emitted from the object positioned at f_2 are reflected twice and finally focused onto the detector placed at the right focus (not shown in this figure) of the ellipsoid mirror.

mirror are denoted by f_1 and f_2, respectively. When x-rays are emitted from point f_2 and propagate in all directions (shown in magenta), following reflection by the hyperboloid mirror, the x-rays propagate as if they originated from f_1. The left focus of the ellipsoid mirror is also located at f_1. The ellipsoid mirror reflects and focuses at its right focus. The Wolter mirror satisfies the Abbe sine condition.

This optic can be used not only to focus primary x-rays to deliver a tiny beam to the sample but also to image x-rays from any point in the sample onto the detector plane. When used for projection-type XRF imaging, the projected image is magnified relative to the object. In astronomy, the Wolter mirror telescope is used to collect parallel x-rays from objects at infinite distances. Therefore, the ellipsoid mirror is replaced by a parabolic mirror whose focus remains at f_2. The detector is placed at the f_2 position.

In general, the use of total reflection x-ray mirrors has inherent challenges because of the relatively small critical angle. Such x-ray mirrors are most likely made of silicon or glass. As shown in figure 3.11, the critical angle of total reflection is less than 5 mrad for x-rays that have energies greater than 6 keV, corresponding to the K-line fluorescence x-rays of chemical elements with atomic numbers greater than 29 (manganese). This small critical angle makes it difficult to design a Wolter mirror with a short focal length. Therefore, to solve this problem, a coating of some metallic layers is desirable. On the other hand, for lower x-ray energies, the problem is somewhat alleviated, and therefore the use of Wolter mirror would be quite reasonable in transmission-type soft x-ray microscopes for biological samples such as cells, although this is not exactly x-ray fluorescence imaging. Figure 3.17 shows a photograph of modern commercially available soft x-ray microscope with two Wolter mirrors [59]. They are made of glass and are applicable to soft x-rays such as 525 and 277 eV. The longer mirror is used

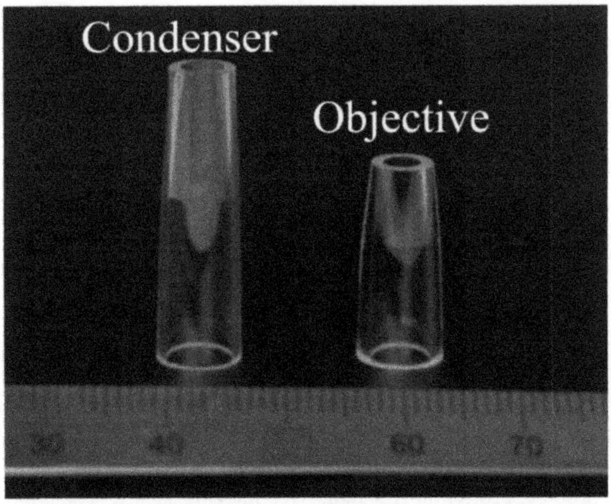

Figure 3.17. A photograph of two glass Wolter mirrors used for a transmission-type soft x-ray microscope. Adapted from figure 2 in reference [59] with permission from AIP Publishing.

to concentrate the primary x-rays that illuminate the object. This means that a micro x-ray source is placed at the right focus of the ellipsoid mirror, while the object is placed at f_2. Its magnification is 1/4.5. The shorter one serves as an objective lens that projects the x-rays transmitted through the object. This means that the object is placed at f_2 and the detector is positioned at the right focus of the ellipsoid mirror. It has a high magnification of 150 and achieves a spatial resolution of 400 nm.

Theoretically, Wolter-type mirrors are the most competitive optics for high-resolution x-ray microscopy. Since 1987, Wolter-type focusing and/or imaging mirrors for hard x-rays have been pioneered by S. Aoki and his colleagues [60–66]. In the 1980s and 1990s, even with synchrotron x-rays, it was not very easy to obtain a small beam for a scanning-type x-ray microscope. With the advent of 3rd-generation synchrotron sources combined with zone plates, the barrier of 1 μm beam size was finally overcome, and it became possible to use a nano-beam even with hard x-rays. A little later, the Kirkpatrick–Baez mirror, which will be discussed in the next section, also broke the 1 μm barrier. At that time, Wolter mirrors were potentially one of the best optics, but the early examples suffered from technical problems [67–72].

Figure 3.18 schematically shows a Wolter mirror developed in 2000 for projection-type XRF imaging [67]. The average angle of incidence was 7 mrad. To reflect x-rays with energies of up to 12 keV, the inner surface of the Wolter mirror was coated with a platinum layer. The distance from the object plane to the image plane was 2200 mm and the imaging magnification was ten. The x-ray detector used had a pixel size of 12 μm, resulting in a theoretical spatial resolution of 1.2 μm. However, due to manufacturing errors in the surface profile of the Wolter mirror,

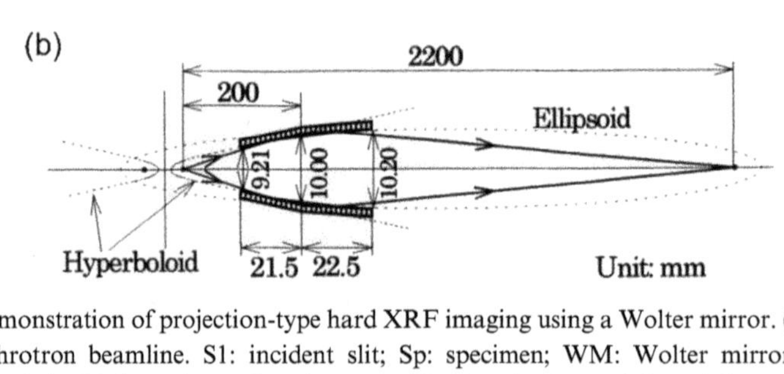

Figure 3.18. A demonstration of projection-type hard XRF imaging using a Wolter mirror. (a) The instrument layout at a synchrotron beamline. S1: incident slit; Sp: specimen; WM: Wolter mirror; CCD: detector. (b) Specifications of the Wolter mirror. Imaging magnification 10 times. Adapted from figures 3 and 5 in reference [67] with permission from AIP Publishing.

the actual spatial resolution did not reach this level. It was estimated to be better than 10 μm, which was inferior to other x-ray optics available at the time for focusing and/or imaging. In the 2020s, many dramatic changes occurred due to advances in both ultralow-emissivity x-ray sources and ultraprecise machining technologies for polishing mirror surfaces. In the not-too-distant future, x-rays may break through the next barrier, at one to a few nanometers. It may now be time for scientists to remember the potential of the Wolter mirrors. Recently, some advanced technology such as organic abrasive machining has been applied to the figure correction of the master mandrel of a Wolter mirror. This resulted in a figure accuracy of <1 nm root mean square by selective removal of a fused silica surface [73]. Wolter-type mirrors with such careful machining may be a promising candidate for future projection-type XRF imaging. One of their major advantages over pinholes, coded apertures, and collimator plates is that a good working distance can be achieved between the sample and the detector, which is convenient for many practical applications.

3.3.5 The Kirkpatrick–Baez mirror

The advent of the Kirkpatrick–Baez (KB) mirror was earlier than that of the Wolter mirror. It was first developed [74] by P Kirkpatrick and A V Baez in 1948 for forming x-ray images. As schematically shown in figure 3.19(a), it consists of a pair of vertical and horizontal mirrors with 1D elliptical surfaces. These two reflecting mirrors focus the x-rays that diverge in the vertical and horizontal directions, respectively. Manufacturing 1D elliptical outside surfaces with an acceptable figure error and surface roughness is much easier than manufacturing 2D curved inner surfaces. The first KB mirror was made of glass coated with a metal film. Nowadays, most KB mirrors are made of single-crystal silicon due to the availability of large, high-purity silicon ingots and their excellent thermal

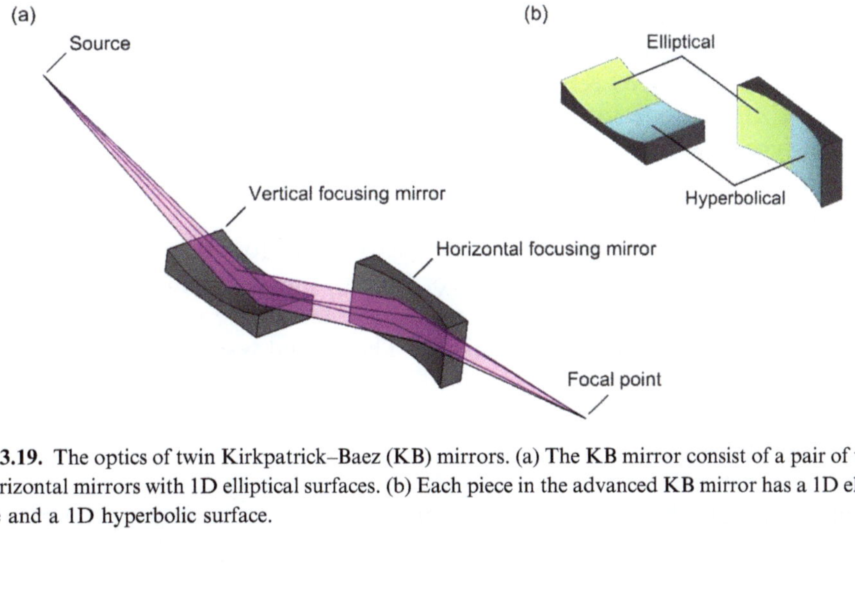

Figure 3.19. The optics of twin Kirkpatrick–Baez (KB) mirrors. (a) The KB mirror consist of a pair of vertical and horizontal mirrors with 1D elliptical surfaces. (b) Each piece in the advanced KB mirror has a 1D elliptical surface and a 1D hyperbolic surface.

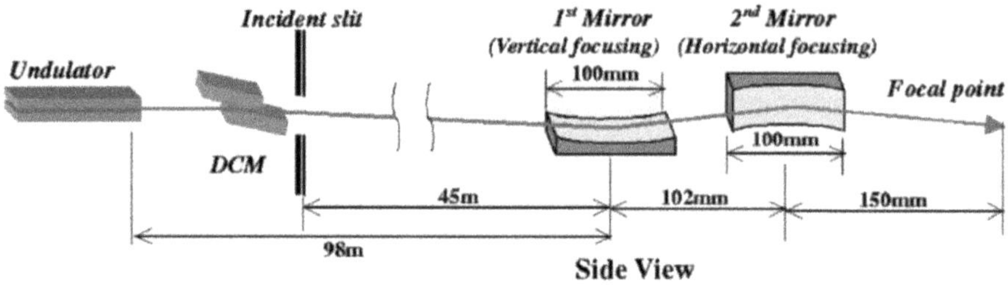

Figure 3.20. An example of a synchrotron beam focused using KB mirrors. Adapted from figure 1 in reference [75] with permission from AIP Publishing.

conductivity. The surface of the KB mirror is often coated with a metal film to increase the critical angle of total reflection.

Although initially developed for forming x-ray images, the KB mirror is mostly employed for focusing synchrotron x-ray beams. This has become a standard in almost every synchrotron facility today. An example [75] is shown in figure 3.20, where the dimensions of the KB mirror and the distance from the mirror to the focus are indicated. The size of the focal point is only 30 nm. The design of the KB mirror, including its surface curvature and focal distance, can be changed depending on the specific requirements.

There have also been some studies (although not many) of the use of KB mirrors in x-ray imaging [76]. However, their elliptical surfaces result in varying incidence angles, leading to comatic aberration and a significantly narrowed field of view. As a result, KB mirrors are not considered well suited for x-ray image projection.

To solve this problem, the advanced KB mirror [77] was developed. As shown in figure 3.19(b), each mirror consists of two parts: a 1D elliptical surface and a 1D hyperbolic surface. In other words, the advanced KB mirror is like a 1D version of the Wolter mirror introduced in section 3.3.4 that replaces the rotational focusing scheme with two perpendicular focusing schemes. Making an advanced KB mirror is more difficult than making a KB mirror. However, the advanced KB mirror almost satisfies Abbe's sine condition, reduces comatic aberration, and has a wide field of view. So far, the advanced KB mirror has succeeded in focusing an x-ray beam to a very small size of about 30 nm, and it also achieved a 50 nm spatial resolution in projection-type x-ray imaging [78]. In this work, the distance from the mirror to the detector was 45 m. The magnifications in the vertical and horizontal directions were 196 and 637, respectively.

Figure 3.21 shows the layout of the instruments for a projection-type XRF imaging experiment utilizing a KB mirror [79]. In this work, the distance from the mirror to the detector was 5.6 m. The magnifications in the vertical and horizontal directions were 25.7 and 83.5, respectively. The spatial resolution produced by the mirror optics was 50 nm, but the actual resolution of the XRF images was 500 nm, which was limited by the pixel size of the detector. Figure 2.18 in chapter 2 was acquired during this experiment.

The advanced KB mirror can be used in the domain of hard x-rays. Due to its high magnification, the sample should be illuminated with high-flux primary x-rays.

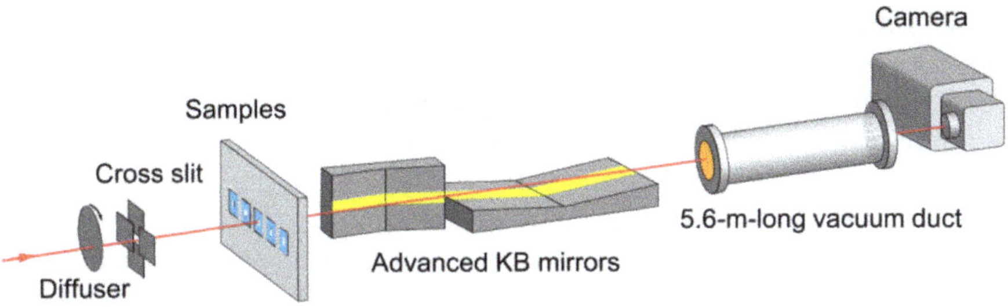

Figure 3.21. The instrument layout of a projection-type XRF imaging experiment utilizing an advanced KB mirror. Adapted with permission from figure 2 in reference [79], copyright 2019 Optical Society of America.

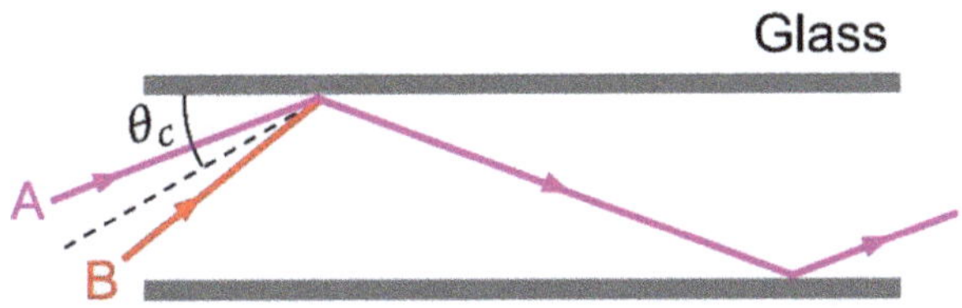

Figure 3.22. The propagation of x-rays inside a hollow glass capillary. θ_c is the critical angle for total reflection at the glass surface.

In addition, due to its long working distance, it is considered to be applicable only in synchrotron facilities. The alignment of the instruments during the experiment is quite challenging and often takes a long time. However, due to its extremely high spatial resolution, which is significantly higher than those of any other optical components used for achromatic x-ray imaging, it can be used to observe fine nanostructures in cutting-edge biological experiments.

3.3.6 Polycapillary optics

A polycapillary optic is composed of a bundle of hollow glass capillaries. These capillaries are separated by walls whose diameters can range from a few to tens of microns. These capillaries can selectively pass incident x-ray photons that propagate nearly parallel to their axes. Polycapillary optics is like a combination of collimation optics and reflection optics. As illustrated in figure 3.22, if the grazing angle of the incident x-ray photon (A) is smaller than the critical angle for total reflection, the incident x-ray photon is reflected and continues to propagate forward within the capillary. In contrast, if the grazing angle is larger than the critical angle (B), the x-ray photon is blocked.

Since they are not limited to projection-type XRF imaging, polycapillary optics are widely used in various x-ray experiments and equipment. Their main benefit is to greatly reduce the inverse-square dependence of the x-ray intensity on the distance from the source. Polycapillary optics can be straight or curved [80]. The design, fabrication, and instrument alignment of the curved capillaries are generally more complicated. Curved polycapillary optics can be used to converge divergent x-rays or collimate divergent x-rays into a parallel beam. The former is applicable when focusing a micro x-ray spot, while the latter can be seen in wavelength-dispersive XRF experiments, where a polycapillary optic is placed between the sample and the crystal analyzer to collect more of the fluorescence x-rays emitted from the sample.

The polycapillary devices used in projection-type XRF imaging are straight, and their optics is relatively simple. It can be divided into two types: parallel polycapillary devices for projecting 1:1 XRF images, and conical polycapillary devices for projecting magnified images [81]. For the former, the back-end diameter d_{out} is the same as the front-end diameter d_{in} of the capillary. For the latter, $d_{out} = Md_{in}$, where M is the imaging magnification factor. As sketched in figure 3.23, let ϕ_c denote the critical angle for total reflection, and let α denote the conical angle. Then, the acceptance angle for incident x-rays is $\alpha_{in} = \phi_c + \alpha/2$. The half-angle of the x-ray divergence cone, α_{out}, lies between half of the conical angle and the acceptance angle: $\alpha/2 < \alpha_{out} < \alpha_{in}$. Let f_1 denote the object–capillary distance and f_2 denote the capillary–image distance. The spatial resolution of a polycapillary optic Δ is given by

$$\Delta = \sqrt{\left(d_{in} + 2\frac{f_2}{M}\tan\alpha_{out}\right)^2 + (2f_1\tan\alpha_{in})^2}\,. \tag{3.10}$$

This equation also applies to parallel polycapillary optics if we set $M = 1$ and $\alpha_{in} = \phi_c$. Therefore, in XRF imaging, it is important to minimize the object–capillary distance f_1 and the capillary–image distance f_2. Typically, f_1 and f_2 are on a scale of several millimeters, and α_{in} and α_{out} are on a scale of several milliradians. As a result, the spatial resolution Δ is on the micrometer scale.

In practical experiments, the spatial resolution is also limited by the detector pixel size. Generally, for parallel polycapillary optics, the spatial resolution is considered equal to the detector pixel size. For conical polycapillary optics, the spatial resolution is estimated by dividing the pixel size by the magnification factor.

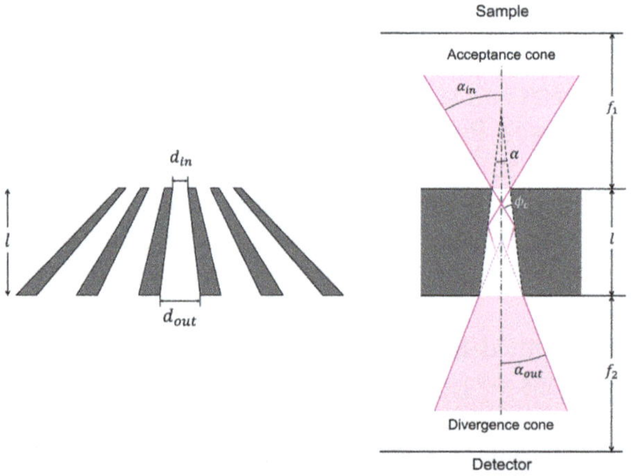

Figure 3.23. Schematic illustration of a conical capillary device and its optics.

With current fabrication technology, magnification factors of up to $M = 10$ can be achieved. In chapter 2, figures 2.3, 2.12, and 2.22 were acquired using a 1:1 parallel polycapillary optic, while figure 2.6 was acquired using a conical polycapillary optic with a magnification factor of six.

For x-rays of different energies, the transmission ratio of polycapillary optics varies due to their different critical angles for total refection at the glass surface. Therefore, when acquiring x-ray images at distinct energies, it is necessary to apply intensity correction for comparison. From this perspective, we can clearly see the difference between the polycapillary and the collimator plate mentioned in section 3.3.3. With the use of the collimator plate, the fraction of x-rays that undergo total reflection is extremely small and negligible. Unlike polycapillary systems, there is no need to consider the different transmission ratios for X-rays of different energies.

Due to the use of the same glass material and similar densely packed porous designs, confusion has arisen in some literature about the naming of polycapillary optics, collimator plates, and microchannel plates. Therefore, the reader should distinguish them using their geometric dimensions and optics. It should be noted that, in addition to the previously introduced collimator plate and polycapillary optics, there is also a unique microchannel plate with square capillaries. It can project XRF images onto a focal plane due to multiple reflections at the corners of the square capillaries. As described in [82], its optics is quite complicated. So far, this squared capillary microchannel plate has not been widely utilized, but figure 2.19 in chapter 2 was acquired using this special optical component.

The specifications of the different x-ray optical components available for projection-type XRF imaging are summarized in table 3.2. Note that the descriptions in the table only represent typical cases seen up to the present. Exceptions always exist, and the authors also anticipate their further development in the future.

3.4 Imaging detectors

3.4.1 Energy-dispersive 2D x-ray detectors

Detectors used for projection-type XRF imaging must meet two requirements: first, they must be able to measure the energy of each x-ray photon, and second, they must be able to record the 2D position of each x-ray photon. The first requirement is the same as for conventional semiconductor x-ray energy-dispersive detectors, as explained in chapter 1.4. The second requirement means that the imaging detector must have a 2D pixel array.

Figure 3.24 shows the working principle of a typical semiconductor energy-dispersive x-ray detector [83]. A reverse bias is applied to a p–n diode. When an x-ray photon is incident on the p–n diode and absorbed, electron–hole pairs are created. The number of electron–hole pairs is proportional to the energy of the x-ray photon, and the coefficient of proportionality depends on the semiconductor material. The most common material for semiconductor detectors is silicon, which requires 3.6 eV to create one electron–hole pair. (For reference, this coefficient is 3 eV for germanium, 5 eV for cadmium zinc telluride, 6.5 eV for mercury iodide, and 5.5 eV for diamond, for more details, see table 1.4.). Thus, a fluorescent x-ray

Table 3.2. Typical specifications of different x-ray optical components.

	Pinhole	Coded aperture	Collimator plate	Wolter mirror	KB mirror	Polycapillary optic
Resolution	20–200 μm	~100 μm	~20 μm	<10 μm	Up to 500 nm	10–50 μm
Magnification	Flexible	Flexible	1:1	10×	20×–700×	1:1 or 3× or 6×
Detector position	Flexible	Flexible	Flexible	Fixed	Fixed	Flexible
Optical length	5–50 mm	5–50 mm	3 mm	~2m	>10 m	~50 mm
Image reconstruction	—	Required	—	—	—	—
X-ray source	Lab or synchrotron	Lab or synchrotron	Lab or synchrotron	Synchrotron recommended	Synchrotron only	Lab or synchrotron
Expense	Very low	Low	Low	Medium	High	Medium

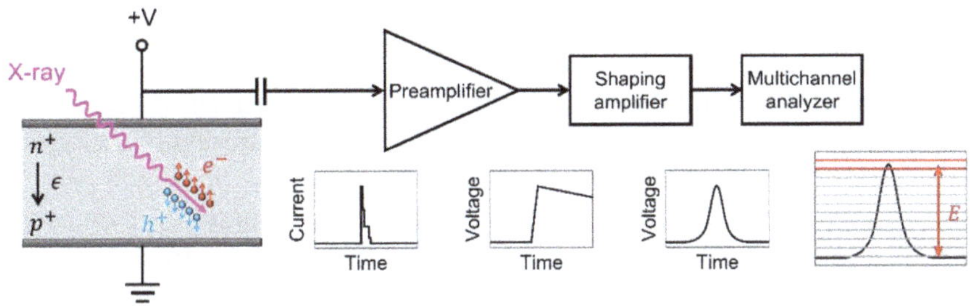

Figure 3.24. The working principle of measuring x-ray photon energy with a semiconductor detector [83].

photon, such as an iron Kα emission at 6.403 keV, can create about 1800 electron–hole pairs in silicon. These charges form a charge cloud in the p–n diode.

The generated electron–hole pairs are separated by a bias (electric field) and drift towards the electrodes at both ends of the p–n junction, where they are collected. As a result, a weak pulsed current is generated in the circuit. This pulsed current is amplified by a preamplifier, resulting in a step-like pulsed voltage. This signal is further processed by a shaping amplifier to produce a quasi-Gaussian voltage pulse. Finally, the amplitude of the voltage pulse is measured by a multichannel analyzer. This amplitude is directly proportional to the energy of the x-ray photon. In other words, after calibration, the energy of the x-ray photon can be measured.

It is important to note that x-rays have a high penetration capability. To maximize the absorption of incident x-ray photons and improve the detection efficiency, the semiconductor p–n diode, or more specifically, the depletion layer, must have a certain thickness. For silicon semiconductor detectors, a thickness of 500 μm or more is required.

In addition to the circuit design of the detector itself, errors are introduced when collecting the charge cloud, amplifying the signal, and measuring the amplitude of the voltage pulse. The magnitude of these errors determines the energy resolution of the detector. Currently, the typical energy resolution of semiconductor x-ray energy detectors [84] is around 130 eV at 5.9 keV (Mn Kα). This is much worse than the energy resolution of wavelength-dispersive XRF analysis, which is typically a few eV. However, it is generally considered sufficient for identifying chemical elements and resolving the chemical composition of common samples.

On the other hand, to record the position of each x-ray photon, the detector must be fabricated as an array of many independent pixels. This greatly increases the complexity of the circuitry and the fabrication process. For silicon x-ray detectors, the fabrication of small pixels is challenging due to the required thickness of the semiconductor layer of several hundred microns. Currently, the smallest pixel size in x-ray detectors is 48 μm. In an attempt to reduce the thickness of the semiconductor layer to facilitate the fabrication of smaller pixels, the use of high-Z semiconductor materials with stronger x-ray absorption capabilities, such as cadmium telluride, is also being considered. The development of x-ray pixel detectors using such high-Z materials is currently underway.

It should be noted that most x-ray pixel detectors to date have been designed, fabricated, and used for x-ray scattering and x-ray diffraction experiments. In these experiments, the primary focus is on the intensity pattern of the x-rays, while information about the energy of each x-ray photon is unnecessary. Conventional x-ray pixel detectors operate in an integration mode, in which the charges produced by many x-ray photons are accumulated in a pixel and then read out. In this mode, the effect of a single high-energy x-ray photon can be similar to that of several low-energy x-ray photons, making it impossible to accurately determine the number of photons. The new generation of x-ray pixel detectors operates in single-photon-counting mode, in which each voltage pulse generated by an x-ray photon is counted. This mode allows the photon counts to be accurately recorded. However, it is still uncommon to use a multichannel analyzer to accurately measure the amplitude of each quasi-Gaussian voltage pulse. Instead, a comparator is introduced: if the pulse amplitude exceeds a certain threshold, it is counted; otherwise, it is not. This threshold can be flexibly adjusted and is often used to filter out low-energy background. Replacing a multichannel analyzer with a comparator reduces the processing time for each photon event and increases the maximum photon counting rate for detectors with millions of pixels, making them suitable for high-intensity x-ray scattering experiments. Currently available single-photon-counting x-ray pixel detectors include Eiger [85], Medipix [86], Pilatus [87], and Epix [88], among others. While they show excellent performance in x-ray detection and have a wide range of applications, they cannot be directly applied to energy-dispersive XRF imaging due to the loss of information about the energy of each x-ray photon. In fact, they were not originally designed for this purpose.

Among these x-ray pixel detectors, the pnCCD may be an exception [89]. It has the ability to measure the energy of each x-ray photon and a relatively high energy resolution of 152 eV at 5.9 keV. This value approaches the typical energy resolution of 130 eV of conventional energy-dispersive x-ray detectors, making the pnCCD suitable for routine energy-dispersive XRF analysis. The pnCCD was originally developed for the astrophysical satellite mission XMM-Newton [90]. Its uniqueness lies in its ability to precisely determine the number of generated charges in each pixel. This is mainly achieved by an on-chip amplifier for initial amplification and a connected multichannel readout application-specific integrated circuit (ASIC) for additional signal amplification, filtering, and multiplexing. The silicon layer of a pnCCD has an effective absorption thickness of 450 μm, resulting in a detection efficiency of over 95% for x-rays in the energy range of 3 to 10 keV and over 30% for 20 keV x-rays. The pixel size of a pnCCD can be as small as 48 μm, and each sensor chip has 264×264 pixels. Its image refresh rate can reach 400 Hz. To date, the pnCCD has been widely used in various fields, including astronomical detection. When combined with a polycapillary optic, it becomes a powerful tool called a color x-ray camera for projection-type XRF imaging. Some examples of pnCCD images are figures 2.4, 2.5, 2.6, and 2.22 shown and explained in chapter 2. One limitation of the pnCCD may be its large pixel size, which may somewhat affect the spatial resolution of the XRF image. More importantly, although the pnCCD has become commercially available as a professional x-ray instrument in recent years, it has not been mass-produced and is only available in cutting-edge x-ray experiments due to its high cost. If the pnCCD were the only detector available for projection-type XRF imaging, the authors believe that most readers of this book would not be able to perform their own experiments.

3.4.2 Charge-coupled devices and complementary metal–oxide–semiconductor cameras for x-ray applications

Fortunately, camera sensors designed for normal visible light (with special operation and image processing described later) can serve as alternatives to the energy-dispersive 2D x-ray detector. Their relatively lower cost, easier accessibility, smaller pixel size, and larger number of pixels make them attractive to general x-ray imaging users. They are still worthwhile even if the detection efficiency of x-ray photons is lower than that of the specially developed energy-dispersive 2D x-ray detectors due to the rather limited thickness of the silicon layer, typically from 10 to 40 μm.

The CCD, first invented by Willard Boyle and George E. Smith at Bell Labs in 1969 [91, 92], is an integrated circuit containing an array of coupled capacitors. As is well known, the device can be used as an image sensor by accumulating photon-generated charges in the depletion layer. The key to the design of the CCD was the ability to transfer charge along the surface of a semiconductor from one storage capacitor to the next. During frame transfer of the image to the controller/computer, the charges stored in each pixel are read line by line. CCD cameras have become very popular and have been used not only for some highly sophisticated scientific/technological applications but also for ordinary personal ubiquitous use in daily life. Willard Boyle and George E. Smith won the Nobel Prize in Physics in 2009 for the invention of an imaging semiconductor circuit - the CCD sensor.

The complementary metal–oxide–semiconductor (CMOS) sensor, which is now even more widely used than the CCD, is a type of active pixel sensor [92]. Compared with the CCD, its fabrication process and signal processing are more advanced, whereas its photon-to-charge conversion scheme is almost the same. To produce CMOS sensors, regular semiconductor fabrication equipment is used. Unlike the CCD, in a CMOS sensor the amount of charge for each pixel is handled by a specific amplification circuit for each pixel, resulting in much faster image transfer. The idea of an active pixel sensor was first proposed by Peter Noble in 1968 [92]. Almost the first practical active pixel sensor, an n-type metal–oxide–semiconductor (NMOS) image sensor, was developed by Olympus in Japan in 1985 [93]. The technology was significantly improved by NASA scientists in 1993 for the space observatory [92]. Nowadays, some CMOS sensors have large pixel sizes and low noise characteristics, as well as more bits of analog-to-digital (A/D) conversion, making them more suitable for scientific application than conventional CMOS. Such sensors are sometimes called sCMOS [94].

CCD and CMOS cameras can also be easily used for x-ray imaging by using a scintillator to convert x-ray photons into visible light. In this case, the sensor is not directly exposed to the incident x-rays. As the x-rays form the visible light image on the scintillator, the images are captured by the camera using optical fibers, mirrors, or lens systems. This approach is widely used in radiography and medical imaging. It is suitable for imaging intense, high-energy x-rays and can provide high-resolution images. However, in the process of converting x-rays into visible light, the information about the energy of each x-ray photon is obviously lost. Therefore, the above scheme cannot be used for x-ray color imaging, i.e. x-ray imaging with energy information, which is essential for XRF imaging.

So, what happens when a sensor is exposed to x-ray photons? The answer is very close to the case of ordinary semiconductor detection. In principle, the semiconductor photosensor can also work for x-rays, but the amount of charge generated is much greater than for visible light. The obvious advantage is that the dark count is much lower than the signal level. However, saturation easily occurs, even with small numbers of x-ray photons per pixel. The actual gradation in the image can be much smaller than the number of levels used for A/D conversion. Another more important point is the information about the x-ray energy. As already explained, the number of charges produced by a single x-ray photon is quite large and proportional to the x-ray energy. Therefore, in this context, the number of charges gives information about the x-ray energy rather than the x-ray intensity. On the other hand, the standard use of this type of camera system is the accumulation of charges to obtain an image. During accumulation, one cannot distinguish between two photon events and a single event with a half-energy x-ray photon. In other words, the information about the energy of the x-ray photons is preserved when a camera is used under single-photon conditions, even though the obtained data may look like a noise image due to low counts. Some pioneering x-ray color imaging based on single-photon counting was studied in x-ray astronomy as early as the 1980s and early 1990s [96–98]. At least since 1998, similar schemes have been introduced in x-ray microscopy by S Aoki [64, 67, 70]. In practice, the frame rate of the camera becomes quite important, because it is necessary to stop recording before two photon events are co-currently recorded in any pixel. If the frame rate is too slow, the measurement may not be practical. Another concern is that the charges generated by a single x-ray photon can appear in multiple pixels. This is called charge sharing [98] and is common to any pixel detector, especially as the pixel size gets smaller. Therefore, it is also important to make corrections for such effects in x-ray color imaging.

3.4.3 Processing for multielement imaging

As mentioned above, there are two technical issues to consider when using semiconductor image sensors for x-ray color imaging. First, it is important that no more than two photon detection events occur in a single pixel. Second, it is important that the results of a single-photon detection event are not distributed over multiple pixels, and that even if they are distributed, reliable correction should be performed. As long as these two conditions are met, it is possible to acquire images while discriminating between x-ray energies using CCD and CMOS cameras that are widely used in visible light applications, as opposed to specially designed and developed 2D x-ray detectors such as pnCCDs.

Since the energy information of x-ray photons is contained in the amount of charge produced in the pixels of the sensor, when two or more photon detection events occur, it is impossible to determine whether the photons have more than twice the energy or twice the number of events. Energy-dispersive x-ray detectors do not have this problem because the detection and readout process for such events is fast. The same is true for pixel detectors for 2D x-rays. On the other hand, in the most common use of commercially available CCD and CMOS cameras for general visible light applications, the charge generated by photon detection events is stored sufficiently to be read out. This is not at all inconvenient in the case of visible light, but it is inconvenient for discriminating between the energies of x-ray photons. Nevertheless, counting single photons is not technically difficult. There is no need to modify the electronic circuitry of the sensor substrate. Single-photon counting means low counts, and simply setting the exposure time appropriately results in conditions where only a few photons are counted in the camera. If a short exposure time still results in a large number of x-ray photons entering the camera, one approach is to attenuate the incident x-rays somewhat. As the reader may have noticed, when using a commercial CCD or CMOS camera under single-photon-counting conditions, the x-ray intensity available from a laboratory x-ray source may be just right. Rather, for practical purposes, it is important that the frame rate of the camera should be high and suitable for frequent readouts.

Figure 3.25 illustrates the difference between accumulation mode and single-photon-counting mode. Accumulation mode is a very common use of commercially available CCD and CMOS cameras for general visible light applications. When imaging is performed using standard camera control software, the camera acquires images in this state; it can also be used as is for x-ray applications, but while an x-ray image is obtained, information about the x-ray energy is lost. The single-photon-counting mode, on the other hand, involves very frequent repeated image acquisitions with extremely short exposure times. It is quite possible to perform this operation using standard camera control software. The choice of exposure time is made after the actual acquisition of a test image in which no more than two photon detection events occur per pixel, so that the acquired image looks like a noise image. In such an image, the x-ray photons are sparsely distributed on the sensor without

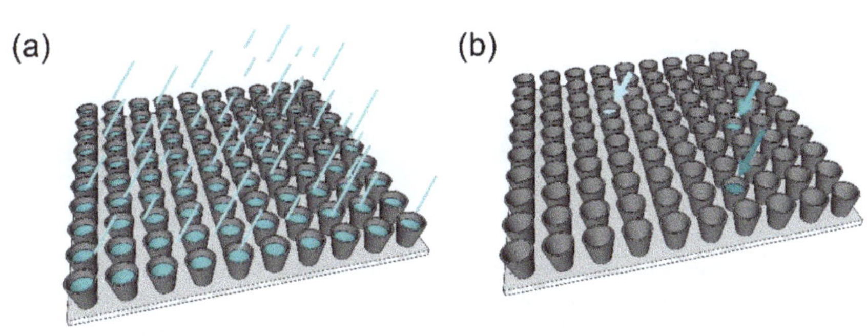

Figure 3.25. Conceptual illustrations of: (a) accumulation mode and (b) single-photon-counting mode

overlapping or connecting. In this way, the energy of each x-ray photon can be detected individually. The resulting image appears noisy and does not contain a continuous pattern. It contains only scattered bright spots, and in principle each bright spot corresponds to an x-ray photon event.

Suppose the pixels at (X_1, Y_1) have a specific digital value corresponding to an x-ray energy E_1, and the pixels at (X_2, Y_2) have a different value corresponding to an x-ray energy E_2. Then, in the preprepared image for x-ray energy E_1, the number of pixels at (X_1, Y_1) should be increased by one. Similarly, the number of pixels at (X_2, Y_2) in the image prepared for x-ray energy E_2 should be increased by one. In this way, many images can be generated for different energies, or, in this case, for different x-ray fluorescence energies, i.e. for different elements. This process can be very similar to calculating the region of interest (ROI) integrated intensity corresponding to each chemical element in the case of quantitative analysis of conventional energy-dispersive x-ray fluorescence spectra.

Even if the image acquisition is done using the single-photon-counting method, there are still some technical problems to solve. It is acceptable as long as all the charges generated by such a photon detection event are stored in a single pixel. Sometimes such charges are distributed over several pixels, which is called the charge-sharing effect. This can happen in any type of 2D semiconductor sensor, especially if the pixel size is small. In this case, it is necessary to make corrections to obtain accurate x-ray energy information for that single-photon event. The authors have investigated three methods [99], namely: (i) integration, (ii) filtering, and (iii) integrated filtering, as shown schematically in figure 3.26.

Charge sharing refers to the situation in which a cloud of charges generated by a single x-ray photon is collected by several adjacent pixels. When all the charges are collected in one pixel, an isolated single signal pixel appears in the image. The single-pixel event is the simplest and ideal case, where the energy of an x-ray photon is proportional to the digital intensity of that pixel. On the other hand, when charge is collected by several adjacent pixels, a collection of adjacent signal pixels appears in the image, which is called a multipixel event. In the ideal case of a multipixel event,

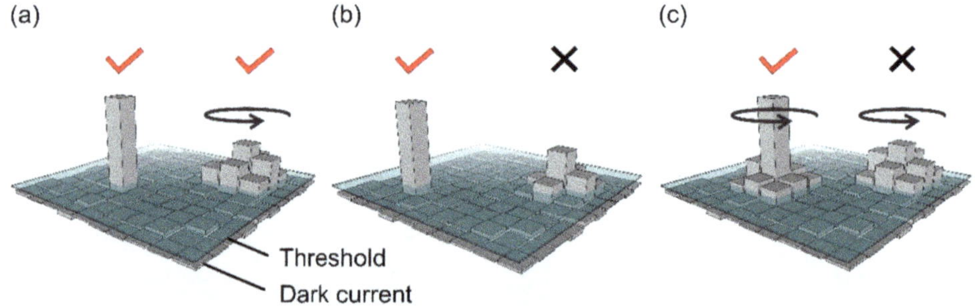

Figure 3.26. Charge-sharing correction strategies. (a) Integration. It accepts single-pixel events and multipixel events. The photon energy of multipixel events is reconstructed, generally in the way of integration. (b) Filtering. It rejects multipixel events and counts single-pixel events only. (c) In integrated filtering, multipixel photon events are classified based on their charge distribution. Only those events with a clear central dominant feature are accepted and then reconstructed.

all charges are fully collected with no charge loss during charge sharing. In this case, the digital intensities of the collected signal pixels are summed, and the energy of the x-ray photons is proportional to the summed intensity. Therefore, image processing must first detect signal pixels from the dark current background of the image and identify whether they are single- or multiple-pixel-events. In the case of a multipixel event, the digital intensities of the clustered signal pixels are summed. This is called the integration method.

It is important to note that even though the number of photon events in each frame is small, there is a small probability that two or more photon events may overlap or connect. Such a phenomenon is similar to the observation of the sum peak in conventional x-ray energy-dispersive detectors. This is a kind of ghost peak that is caused simply by counting two events as a single event. Such low-probability events can be left alone; otherwise, one can examine the 2D charge distribution for each multipixel event: if the 2D charge distribution resembles a 2D Gaussian distribution with only one high-brightness center, then this multipixel event corresponds to only one x-ray photon, whose position should be the gravitational center of the distributed region. In general, there is a high probability that it contains only one x-ray photon. If the shape of the distribution is strange, there may be more than one x-ray photon present. Such multipixel events should be rejected.

Obviously, the ideal case is to have only a single pixel event for a single x-ray photon. Since, in many cases, charge sharing occurs with little or no charge loss, image processing would only require the integration method. However, as is the case with a specific type of visible light camera sensor, charge loss occurs during charge sharing, and the sum of the digital intensities of several adjacent pixels may not allow the extraction of the energy of the x-ray photons. In such cases, if a significant number of single-pixel events remain in the image, only the single-pixel events should be accepted and all multiple-pixel events should be rejected. This process is called the filtering method. Although the filtering method reduces the efficiency of x-ray detection, it often improves the quality of the XRF spectra, for example by improving energy resolution and reducing the spectral peak tail on the low-energy side.

How should the images be processed when all photon events in an image are multipixel events and no single-pixel events are present? This may happen when using the latest high-resolution cameras with a tiny pixel size. In this case, it would be necessary to examine the 2D charge distribution within each multipixel event to infer whether or not charge loss has occurred during charge sharing. If the charge remains concentrated in the central pixel of the multipixel event, it is a valid and acceptable photon event and the digital intensities of all adjacent signal pixels should just be summed. On the other hand, if the charge distribution is very diffuse, some kind of inelastic process should be suspected, such as the absorption of x-ray photons outside the effective depletion layer of the p–n diode, resulting in a diffuse charge cloud that is not fully collected. This is not a valid photon event and should be rejected. This scheme is called integral filtering.

Even when we consider all the above cases, the image processing is quite straightforward. Although the number of raw data images is huge, the algorithm for this image processing is very simple. You can either write your own simple short software code or use popular free or paid image-processing tools available on the market. The macro function of the software that comes with the camera is also a promising option.

3.5 Building a homemade x-ray color imager

An important and interesting question is how to develop an experimental setup in any laboratory/workplace for x-ray color imaging that can simultaneously acquire x-ray fluorescence images for many chemical elements contained in the samples. In short, is it possible to build such a x-ray color imager by the do-it-yourself (DIY) method at a reasonable cost? The answer is 'yes.'

Figure 3.27 shows a good example built by the authors in their laboratory [100]. Anyone could obtain a similar one by assembling commercially available items and making minor modifications. Is there unused x-ray diffraction equipment in the materials analysis center of your institution or elsewhere? The x-ray source part of such an x-ray diffractometer can be used as is. In our case, we used a 1.5 kW copper x-ray tube for x-ray diffraction. This is very standard and can be found anywhere. The apparatus came with a highly oriented pyrolytic graphite (HOPG) crystal monochromator, which was used without modification. Some newer products may come with a multilayer mirror. However, it does not matter. Even if a monochromator is not included, it is not a problem, but the x-ray beam size used is narrow in one direction and long in the other direction, so a monochromator is a good match. So, if you have one, use it as is. On the other hand, of course, other laboratory x-ray sources can be used for x-ray imaging, x-ray fluorescence, and other x-ray experiments. Even a miniature, low-power (1–50 W) source will work well, since the main point is the x-ray intensity at the sample position. The efficiency of the geometry is important.

For the 2D x-ray detector, as mentioned in the previous section, commercially available CCD and/or CMOS cameras designed for visible light can be used. Table 3.3 shows an example of the specifications of the cameras used in the authors' laboratory. In order to use them for x-ray color imaging, it is necessary to modify the front cover of the camera. More precisely, it is recommended to replace the front cover with the new one prepared for x-ray experiments. Figure 3.28 shows such a modification. In the authors' laboratory, the camera was initially used for visible light imaging. Therefore, it was equipped with several lenses. For x-ray experiments, all of these should be removed. The glass plate should also be removed. Instead, an x-ray window is needed. Beryllium would be a promising material; however, the use of aluminum-coated Mylar film can be a much cheaper alternative if the target elements are not low-Z elements. It is necessary to carefully select suitable sealing and adhesive materials for mounting the x-ray window. It is important to prevent the ingress of visible light. If a collimator plate is used as the imaging optics, the extremely tight geometry is key. Therefore, the new front cover must accommodate the collimator within the x-ray window. The collimator plate is positioned very close to the sensor surface. On the other hand, a micro pinhole can be placed outside the

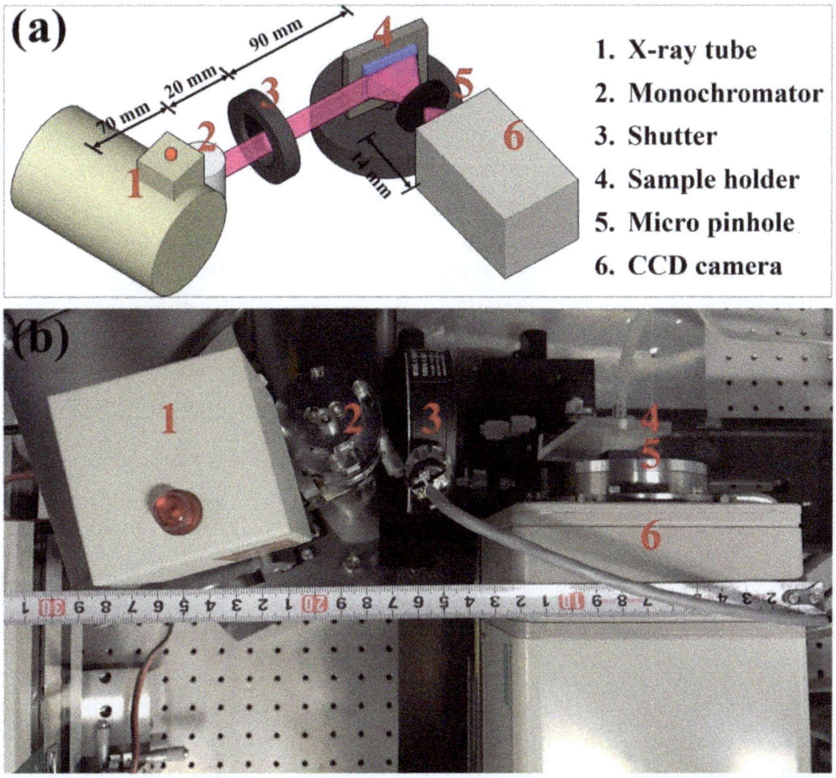

Figure 3.27. An example of an experimental setup for x-ray color imaging based on a commercially available CCD camera. Reprinted from [100], with permission from AIP Publishing. (a) Schematic representation, (b) top view photograph. In this case, a 1.5 kW x-ray tube (Toshiba), a highly oriented pyrolytic graphite (HOPG) monochromator, and a CCD camera (Type I in table 3.3) were used. The beam size at the specimen position was 1 mm (H) × 5 mm (V). The width of the illuminated area of the sample was controlled by adjusting the glancing angle. The front cover of the CCD camera was designed and manufactured in the laboratory. The camera window was made of aluminum-coated Mylar film, which can block visible light but allows x-ray observation. This front cover can be used to store the collimator plate inside the camera but can be left empty when a micro pinhole is used for imaging. A micro pinhole is glued onto the surface of the front cover. If both the collimator plate and the pinhole are removed, this setup can then be used as an energy-dispersive XRF spectrometer with a large-area detector. The distance from the sample holder to the pinhole is not fixed. The dimensions of the CCD camera are 220 mm (L) × 120 mm (W) × 120 mm (H).

x-ray window. Since the sensor must be cooled to reduce electronic noise, the image sensor can quickly freeze and be damaged if water vapor is present in the atmosphere. Therefore, it is important to maintain a vacuum around the image sensor. Some scientific CMOS cameras are designed to operate at low temperatures such as 5 °C, and their image sensors can be kept in a protective atmosphere such as dry nitrogen gas. Overall, the image sensor housing should have excellent airtightness to prevent air leakage.

The sample cell/support is another important item that must be designed and fabricated specifically for the laboratory/workplace. It depends on the objective and

Table 3.3. Example specifications of commercially available cameras which can be used for x-ray fluorescence movie applications.

	Type I	Type II
Sensor	CCD47-10 (e2V) CCD	CIS2521 (Fairchild Imaging) Scientific CMOS
Resolution	1024×1024 pixels	2560×2120 pixels
Pixel size	13 μm × 13 μm	6.5 μm × 6.5 μm
Active area	~170 mm^2	~230 mm^2
Effective thickness	~10 μm	<10 μm
Camera	C4880-50 (Hamamatsu)	pco.edge 5.5 (PCO AG)
A/D conversion	16 bits	16 bits
Electronic cooling	−30 °C with water cooling	5 °C with air cooling
Frame rate (slow scan for high-quality imaging)	0.25 fps (slow mode)	33.6 fps
Energy resolution for x-rays (@ MnKα—5.9 keV)	150 eV	220 eV

application of the observation. However, x-ray imaging is very flexible and can be easily adapted to many different types of experiments by modifying the sample cell/support. In the authors' laboratory, an electrochemical sample cell/support [101] has been used frequently. The cell/support is equipped with x–y–z and θ-ψ mechanical stages for linear and angular movement.

At the beginning of an x-ray experiment, it is crucial to carefully analyze the camera images acquired in single-photon-counting mode to determine the appropriate methods of image processing, especially for charge-sharing correction. When acquiring camera images in single-photon-counting mode, it is essential to coordinate the x-ray intensity and camera exposure time to ensure a relatively low probability of x-ray photon events overlapping. In an XRF imaging experiment, it is often necessary to acquire thousands or tens of thousands of single-photon-counting camera images. If the digital camera used offers a software development kit, users can implement their own code to perform real-time processing of the image data transferred in the buffer. This eliminates the need to store a large number of images on the hard disk, which often hinders the efficiency of image acquisition and continuous image acquisition.

Even before mounting the imaging optics, it is extremely useful to test the camera as an energy-dispersive x-ray spectrometer. Such an example is shown in figure 3.29 [99]. The Type II camera in table 3.3 can be used for x-ray fluorescence analysis, although the camera is typically used for visible light applications. The camera can resolve x-ray spectra and distinguish the presence of cobalt, which is contained only on the front side of the dish plate.

Figure 3.28. Modifying an ordinary visible light camera (Type II in table 3.3) for x-ray fluorescence imaging. Remove all lenses and any glass plates. An aluminum alloy aperture plate is mounted. An aluminum-coated Mylar film is glued to the other side of this plate. On the front, the aperture is covered by tungsten foil with a micro pinhole. When a collimator plate is used as the imaging optic, a different front cover for the camera is required to accommodate the collimator plate at a position extremely close to the sensor surface. In this case, the aperture shape and size are the same as the sensor.

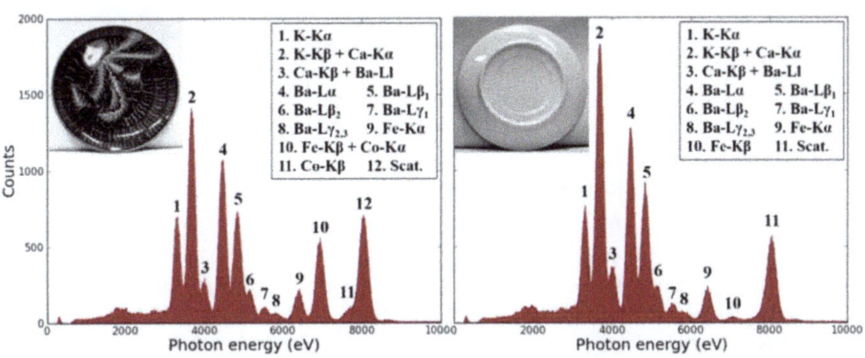

Figure 3.29. Testing the capability of an ordinary visible light camera (Type II in table 3.3) as an energy-dispersive x-ray spectrometer. Reproduced from [99]. CC BY 4.0. Copyright 2017, The Author(s). The sample used is a ceramic plate. The photos shown as insets were taken with the same camera and lenses. The x-ray fluorescence spectra were obtained for the front (left) and back (right) of the plate by simply switching to the x-ray window and x-ray imaging optics (a micro pinhole in this case). The spectra are those of the chemical elements contained in the plate. Some differences can be observed between the front and back. Cobalt was detected in the front of the plate, corresponding to a blue pigment.

Here is a summary for those who want to build a x-ray color imager for their own use.

1. An ordinary x-ray tube can be used for x-ray color imaging. The use of a synchrotron beamline is a possible choice for advanced analysis but is not essential for x-ray color imaging of chemical elements. Laboratory and even mobile x-ray tubes can be used.
2. The instrumental difference between x-ray color imaging and conventional energy-dispersive x-ray fluorescence analysis is only the detector. The use of a 2D detector equipped with imaging optics is the main point.
3. Commercially available CCD and CMOS sensors can be used as 2D detectors for x-ray color imaging. They are usually inexpensive and can be purchased by anyone.
4. Simple, inexpensive imaging optics such as a pinhole or collimator plate are easy for anyone to make. X-ray color imaging can be done by simply placing a pinhole on the 2D detector surface or a collimator plate inside the 2D detector. Both are also commercially available at low cost.
5. The keys to the measurement are that the data should be collected under single-photon-counting conditions and that a correction is required for the charge-sharing events. Note that the number of images will be huge. However, such processing can be done automatically and effectively by simple software.

X-ray color imaging is an extension of x-ray fluorescence analysis. It provides snapshot and movie images in addition to the usual analysis of average chemical composition. Therefore, it is backward compatible with conventional energy-dispersive x-ray fluorescence analysis. X-ray color imaging is not difficult for anyone to use anywhere. If the reader already has x-ray tubes or an x-ray fluorescence spectrometer, then the reader can do x-ray color imaging.

References

[1] Schwinger J 1949 On the classical radiation of accelerated electrons *Phys. Rev.* **75** 1912 https://doi.org/10.1103/PhysRev.75.1912

[2] Feynman R, Leighton R and Sands M 2011 Relativistic effects in radiation *The Feynman Lectures on Physics* **vol I** 50th new millennium edn (New York, NY: Basic Books) 34 https://feynmanlectures.caltech.edu/I_34.html

[3] Winick H 1980 Properties of synchrotron radiation in *Synchrotron Radiation Research* ed H Winick and S Doniach (New York: Plenum) ch 2 https://doi.org/10.1007/978-1-4615-7998-4

[4] Talman R 2006 *Accelerator X-Ray Sources* (New York: Wiley) https://doi.org/10.1002/9783527610303

[5] Mills D M (ed) *'Third-Generation Hard X-Ray Synchrotron Radiation Sources: Source Properties, Optics, and Experimental Techniques'* (New York: Wiley) 2008

[6] Wiedemann H 2010 *'Synchrotron Radiation'* (Berlin: Springer) https://doi.org/10.1007/978-3-662–05312-6

[7] Als-Nielsen J and McMorrow D 2011 Sources *Elements of Modern X-Ray Physics* 2nd edn (New York: Wiley) 2 https://doi.org/10.1002/9781119998365
[8] Willmott P 2019 Synchrotron physics *An Introduction to Synchrotron Radiation: Techniques and Applications* 2nd edn (New York: Wiley) 3 https://doi.org/10.1002/9781119280453
[9] Zhao W, Hirano K and Sakurai K 2019 Expanding a polarized synchrotron beam for full-field x-ray fluorescence imaging *Rev. Sci. Instrum.* **90** 113704 https://doi.org/10.1063/1.5115421
[10] Beckhoff E B, Kanngieser B, Langhoff N, Wedell R and Wolff H 2006 *Handbook of Practical X-Ray Fluorescence Analysis* (Berlin: Springer) https://doi.org/10.1007/978-3-540-36722-2
[11] https://crtsite.com/page5.html
[12] Poludniowski G, Omar A, Bujila R and Andreo P 2021 Technical note: SpekPy v2.0—a software toolkit for modeling x-ray tube spectra *Med. Phys.* **48** 3630–7 https://doi.org/10.1002/mp.14945
[13] Taylor A 1949 A 5 kW. Crystallographic x-ray tube with a rotating anode *J. Sci. Instrum.* **26** 225 https://doi.org/10.1088/0950-7671/26/7/301
[14] Davies D A, Mathieson A M L and Stiff G M 1959 Rotating-anode x-ray generator *Rev. Sci. Instrum.* **30** 488–91 https://doi.org/10.1063/1.1716662
[15] Yoneda Y, Horiuchi T, Fukumori T and Ando H 1976 An x-ray generator with high speed rotating anode *Oyo Butsuri (Appl Phys)* **45** 523–7 (in Japanese) https://doi.org/10.11470/oubutsu1932.45.523
[16] Sakurai K, Osaka N, Sakurai H and Izawa H 1997 New rotating anode x-ray generator for XAFS experiments *Adv. X-Ray Anal.* **39** 149–53 https://doi.org/10.1154/S0376030800022552
[17] Sakurai K 1992 EXAFS experiments with high-power rotating anode x-rays *Jpn. J. Appl. Phys.* **32** 261 https://doi.org/10.7567/JJAPS.32S2.261
[18] Hemberg O, Otendal M and Hertz H M 2003 Liquid-metal-jet anode electron-impact x-ray source *Appl. Phys. Lett.* **83** 1483–5 https://doi.org/10.1063/1.1602157
[19] Otendal M, Tuohimaa T, Vogt U and Hertz H M 2008 A 9keV electron-impact liquid-gallium-jet x-ray source *Rev. Sci. Instrum.* **79** 016102 https://doi.org/10.1063/1.2833838
[20] Larsson D H, Takman P A C, Lundström U, Burvall A and Hertz H M 2011 A 24 keV liquid-metal-jet x-ray source for biomedical applications *Rev. Sci. Instrum.* **82** 123701 https://doi.org/10.1063/1.3664870
[21] 2008 *Portable X-Ray Fluorescence Spectrometry: Capabilities For In Situ Analysis* ed P J Potts and M West (London: RSC Publishing) https://doi.org/10.1039/9781847558640
[22] Mini-X2 X-Ray Tube https://amptek.com/products/mini-x2-x-ray-tube
[23] Mini Focus Packaged X-ray Tube https://microxray.com/products/mini-focus-packaged-x-ray-tube/
[24] Spielmann C 2004 Laser-driven x-ray sources *X-Ray Spectrometry: Recent Technological Advances* ed K Tsuji, J Injuk and R Van Grieken (New York: Wiley) ch 2.3 https://doi.org/10.1002/0470020431
[25] Key M H 1980 Spectroscopy of laser-produced plasmas *Adv. At. Mol. Phys.* **16** 201–80 https://doi.org/10.1016/S0065-2199(08)60009-3
[26] Corde S, Ta Phuoc K, Lambert G, Fitour R, Malka V, Rousse A, Beck A and Lefebvre E 2013 Femtosecond x-rays from laser-plasma accelerators *Rev. Mod. Phys.* **85** 1 https://doi.org/10.1103/RevModPhys.85.1

[27] Miaja-Avila L *et al* 2016 Ultrafast time-resolved hard x-ray emission spectroscopy on a tabletop *Phys. Rev.* X **6** 031047 https://doi.org/10.1103/PhysRevX.6.031047
[28] Milburn R H 1963 Electron scattering by an intense polarized photon field *Phys. Rev. Lett.* **10** 75 https://doi.org/10.1103/PhysRevLett.10.75
[29] Du Y, Yan L, Hua J, Du Q, Zhang Z, Li R, Qian H, Huang W, Chen H and Tang C 2013 Generation of first hard x-ray pulse at Tsinghua Thomson scattering x-ray source *Rev. Sci. Instrum.* **84** 053301 https://doi.org/10.1063/1.4803671
[30] https://nobelprize.org/prizes/physics/2023/summary/
[31] Krause J L, Schafer K J and Kulander K C 1992 High-order harmonic generation from atoms and ions in the high intensity regime *Phys. Rev. Lett.* **68** 3535 https://doi.org/10.1103/PhysRevLett.68.3535
[32] Spielmann C, Burnett N H, Sartania S, Koppitsch R, Schnerer M, Kan C, Lenzner M, Wobrauschek P and Krausz F 1997 Generation of coherent x-rays in the water window using 5-femtosecond laser pulses *Science* **278** 6614 https://doi.org/10.1126/science.278.5338.661
[33] Popmintchev T *et al* 2012 Bright coherent ultrahigh harmonics in the keV x-ray regime from mid-infrared femtosecond lasers *Science* **336** 1287–91 https://doi.org/10.1126/science.1218497
[34] https://slac.stanford.edu/pubs/slacpubs/15000/slac-pub-15225.pdf https://www-ssrl.slac.stanford.edu/lcls/commissioning/documents/th3pbi01.pdf
[35] Refraction and Reflection from Interfaces 2011 *Elements of Modern X-Ray Physics* (New York: Wiley) pp 69–112 https://doi.org/10.1002/9781119998365.ch3
[36] Sanli U T *et al* 2018 3D nanofabrication of high-resolution multilayer fresnel zone plates *Adv. Sci.* **5** 1800346 https://doi.org/10.1002/advs.201800346
[37] Bajt S *et al* 2018 X-ray focusing with efficient high-NA multilayer Laue lenses *Light: Sci. Appl.* **7** 17162 https://doi.org/10.1038/lsa.2017.162
[38] Brewster D 1856 *The Stereoscope: Its History, Theory, and Construction, with Its Application* (Wentworth Press)
[39] Zhao W and Sakurai K 2019 Multi-element x-ray movie imaging with a visible-light CMOS camera *J. Synchrotron Radiat.* **26** 230–3 https://doi.org/10.1107/S1600577518014273
[40] Dicke R H 1968 Scatter-hole cameras for x-rays and gamma rays *Astrophys. J.* **153** L101–6 https://doi.org/10.1086/180230
[41] Ables J G 1968 Fourier transform photography: a new method for x-ray astronomy *Publ. Astron. Soc. Aust.* **1** 172–3 https://doi.org/10.1017/S1323358000011292
[42] Caroli E, Stephen J B, Di Cocco G, Natalucci L and Spizzichino A 1987 Coded aperture imaging in x- and gamma-ray astronomy *Space Sci. Rev.* **45** 349–403 https://doi.org/10.1007/BF00171998
[43] Cieślak M J, Gamage K A A and Glover R 2016 Coded-aperture imaging systems: past, present and future development—a review *Radiat. Meas.* **92** 59–71 https://doi.org/10.1016/j.radmeas.2016.08.002
[44] Haboub A, MacDowell A A, Marchesini S and Parkinson D Y 2014 Coded aperture imaging for fluorescent x-rays *Rev. Sci. Instrum.* **85** 063704 https://doi.org/10.1063/1.4882337
[45] Kulow A, Buzanich A G, Reinholz U, Streli C and Radtke M 2020 On the way to full-field x-ray fluorescence spectroscopy imaging with coded apertures *J. Anal. At. Spectrom.* **35** 347–56 https://doi.org/10.1039/c9ja00232d

[46] Kulow A, Buzanich A G, Reinholz U, Emmerling F, Hampel S, Fittschen U E A, Streli C and Radtke M 2020 Comparison of three reconstruction methods based on deconvolution, iterative algorithm and neural network for x-ray fluorescence imaging with coded aperture optics *J. Anal. At. Spectrom.* **35** 1423–34 https://doi.org/10.1039/d0ja00146e

[47] Siddons D P *et al* 2020 A coded aperture microscope for x-ray fluorescence full-field imaging *J. Synchrotron Rad.* **27** 1703–6 https://doi.org/10.1107/S1600577520012308

[48] Soltau J, Meyer P, Hartmann R, Strüder L, Soltau H and Saldi T 2023 Full-field x-ray fluorescence imaging using a Fresnel zone plate coded aperture *Optica* **10** 127–33 https://doi.org/10.1364/OPTICA.477809

[49] Wiza J L 1979 Microchannel plate detectors *Nucl. Instrum. Methods* **162** 587–601 https://doi.org/10.1016/0029-554X(79)90734-1

[50] Yamaguchi N, Aoki S and Miyoshi S 1987 Two-dimensional imaging x-ray spectrometer using channel-plate collimator *Rev. Sci. Instrum.* **58** 43–4 https://doi.org/10.1063/1.1139564

[51] Koike M, Suzuki I H, Nakae S and Hasegawa K 1990 A 2D imaging detector system for x-ray fluorescence analysis *Nucl. Instr. Meth. Phys. Res.* **A299** 567–70 https://doi.org/10.1016/0168-9002(90)90845-W

[52] Aota T, Yamaguchi N, Ikeda K, Aoki S, Yoshikawa M, Mase A and Tamano T 1997 Image characteristics of a channel plate collimator for low energy x-rays *Rev. Sci. Instrum.* **68** 1661–7 https://doi.org/10.1063/1.1147974

[53] Wroblewski T, Geier S, Hessmer R, Schreck M and Rauschenbach B 1995 X-ray imaging of polycrystalline materials *Rev. Sci. Instrum.* **66** 3560–2 https://doi.org/10.1063/1.1145469

[54] Wroblewski T 1996 An x-ray camera *Synchrotron Radiat. News* **9** 14 https://doi.org/10.1080/08940889608602867

[55] Sakurai K 1999 Total-reflection x-ray fluorescence imaging *Spectrochim. Acta, Part B At. Spectrosc.* **54** 1497–503 https://doi.org/10.1016/S05848547(99)00071-3

[56] Sakurai K and Eba H 2003 Micro x-ray fluorescence imaging without scans: toward an element-selective movie *Anal. Chem.* **75** 355–9 https://doi.org/10.1021/ac025793h

[57] Sakurai K and Mizusawa M 2004 Quick atomic-scale structure imaging by synchrotron x-rays: A new tool for probing realistic inhomogeneous systems *Nanotechnology* **15** S428 https://doi.org/10.1088/09574484/15/6/021

[58] Wolter H 1952 Spiegelsysteme streifenden Einfalls als abbildende Optiken für Röntgenstrahlen *Ann. Phys.* **445** 94–114 https://doi.org/10.1002/andp.19524450108

[59] Ohba A, Nakano T, Onoda S, Mochizuki T, Nakamoto K and Hotaka H 2021 Laboratory-size x-ray microscope using wolter mirror optics and an electron-impact x-ray source *Rev. Sci. Instrum.* **92** 093704 https://doi.org/10.1063/5.0059906

[60] Aoki S *et al* 1987 Imaging characteristics of a replicated Wolter type. I. X-ray mirror designed for laser plasma diagnostics *Jpn. J. Appl. Phys.* **26** 952 https://doi.org/10.1143/JJAP.26.952

[61] Gohshi Y, Aoki S, Iida A, Hayakawa S, Yamaj H and Sakurai K 1987 A scanning x-ray fluorescence microprobe with synchrotron radiation *Jpn. J. Appl. Phys.* **26** L1260 https://doi.org/10.1143/JJAP.26.L1260

[62] Aoki S, Ogata T, Sudo S and Onuki T 1992 Sub-100 nm-resolution grazing incidence soft x-ray microscope with a laser-produced plasma source *Jpn. J. Appl. Phys.* **31** 3477 https://doi.org/10.1143/JJAP.31.3477

[63] Onuki T, Sugisaki K and Aoki S 1992) Fabrication of Wolter type I mirror for soft x-ray *Proc. SPIE 1720, 258. Int. Symp. on Optical Fabrication, Testing, and Surface Evaluation* https://doi.org/10.1117/12.132133

[64] Aoki S, Takeuchi A and Ando M 1998 Imaging x-ray fluorescence microscope with a Wolter-type grazing-incidence mirror *J. Synchrotron Radiat.* **5** 1117–8 https://doi.org/10.1107/S0909049597018542

[65] Takano H, Yokota K and Aoki S 1999 Soft x-ray dark-field imaging microscope with Wolter-type grazing-incidence mirrors *Jpn. J. Appl. Phys.* **38** L1485 https://doi.org/10.1143/JJAP.38.L1485

[66] Yamamoto K, Watanabe N, Takeuchi A, Takano H, Aota T, Fukuda M and Aoki S 2000 Mapping of a particular element using an absorption edge with an x-ray fluorescence imaging microscope *J. Synchrotron Rad.* **7** 34–9 https://doi.org/10.1107/S0909049599014260

[67] Takeuchi A, Aoki S, Yamamoto K, Takano H, Watanabe N and Ando M 2000 Full-field x-ray fluorescence imaging microscope with a Wolter mirror *Rev. Sci. Instrum.* **71** 1279–85 https://doi.org/10.1063/1.1150454

[68] Watanabe N, Yamamoto K, Takano H, Ohigashi T, Yokosuka H, Aota T and Aoki S 2001 X-ray fluorescence microtomography with a Wolter mirror system *Nucl. Instrum. Meth. Phys. Res* **A467–468** 837–40 https://doi.org/10.1016/S01689002(01)00476-4

[69] Takano H *et al* 2002 X-ray scattering microscope with a Wolter mirror *Rev. Sci. Instrum.* **73** 2629–33 https://doi.org/10.1063/1.1487888

[70] Hoshino M, Ishino T, Namiki T, Yamada N, Watanabe N and Aoki S 2007 Application of a charge-coupled device photon counting technique to three-dimensional element analysis of a plant seed (alfalfa) using a full-field x-ray fluorescence imaging microscope *Rev. Sci. Instrum.* **78** 073706 https://doi.org/10.1063/1.2756632

[71] Hoshino M and Aoki S 2008 Laboratory-scale soft x-ray imaging microtomography using Wolter mirror optics *Appl. Phys. Express* **1** 067005 https://doi.org/10.1143/APEX.1.067005

[72] Aoki S, Watanabe N, Asami H and Shimada A 2016 Design of axisymmetric multi-mirror grazing incidence system to increase the numerical aperture of neutron and x-ray microscopes *Opt. Rev.* **23** 161–71 https://doi.org/10.1007/s10043-016-0191-0

[73] Egawa S *et al* 2023 Figure correction of a Wolter mirror master mandrel by organic abrasive machining *Rev. Sci. Instrum.* **94** 053707 https://doi.org/10.1063/5.0145122

[74] Kirkpatrick P and Baez A V 1948 Formation of optical images by x-rays *J. Opt. Soc. Am.* **38** 766–74 https://doi.org/10.1364/JOSA.38.000766

[75] Matsuyama S, Mimura H, Yumoto H, Sano Y, Yamamura K, Yabashi M, Nishino Y, Tamasaku K, Ishikawa T and Yamauchi K 2006 Development of scanning x-ray fluorescence microscope with spatial resolution of 30 nm using Kirkpatrick–Baez mirror optics *Rev. Sci. Instrum.* **77** 103102 https://doi.org/10.1063/1.2358699

[76] Marshall F J and Su Q 1995 Quantitative measurements with x-ray microscopes in laser-fusion experiments *Rev. Sci. Instrum.* **66** 725–7 https://doi.org/10.1063/1.1146270

[77] Kodama R, Katori Y, Iwai T, Ikeda N, Kato Y and Takeshi K 1996 Development of an advanced Kirkpatrick–Baez microscope *Opt. Lett.* **21** 1321 https://doi.org/10.1364/ol.21.001321

[78] Matsuyama S, Yasuda S, Yamada J, Okada H, Kohmura Y, Yabashi M, Ishikawa T and Yamauchi K 2017 50-nm-Resolution full-field x-ray microscope without chromatic

aberration using total-reflection imaging mirrors *Sci. Rep.* **7** 46358 https://doi.org/10.1038/srep46358

[79] Matsuyama S, Yamada J, Kohmura Y, Yabashi M, Ishikawa T and Yamauchi K 2019 Full-field x-ray fluorescence microscope based on total-reflection advanced Kirkpatrick–Baez mirror optics *Opt. Express* **27** 18318 https://doi.org/10.1364/OE.27.018318

[80] Schields P J, Gibson D M, Gibson W M, Gao N, Huang H and Ponomarev I Y 2002 Overview of polycapillary x-ray optics *Powder Diffr.* **17** 70–80 https://doi.org/10.1154/1.1482080

[81] Nowak S H *et al* 2017 Road to Micron Resolution with a Color X-Ray Camera—Polycapillary Optics Characterization arXiv.1705.08939

[82] Chapman H N, Nugent K A and Wilkins S W 1991 X-ray focusing using square channel-capillary arrays *Rev. Sci. Instrum.* **62** 1542–61 https://doi.org/10.1063/1.1142432

[83] Beckhoff B, Kanngießer B, Langhoff N, Wedell R and Wolff H 2007 X-ray detectors and signal processing *Handbook of Practical X-Ray Fluorescence Analysis* (Berlin: Springer Science & Business Media) pp 203–61 https://doi.org/10.1007/978-3-540-36722-2_4

[84] Lechner P *et al* 2001 Silicon drift detectors for high count rate x-ray spectroscopy at room temperature *Nucl. Instrum. Methods Phys. Res. Sect. A Accel. Spectrometers, Detect. Assoc. Equip.* **458** 281–7 https://doi.org/10.1016/S01689002(00)00872-X

[85] Dinapoli R *et al* 2011 EIGER: Next generation single photon counting detector for x-ray applications *Nucl. Instrum. Methods Phys. Res.* A **650** 79–83 https://doi.org/10.1016/j.nima.2010.12.005

[86] Ballabriga R, Campbell M, Heijne E, Llopart X, Tlustos L and Wong W 2011 Medipix3: A 64 k pixel detector readout chip working in single photon counting mode with improved spectrometric performance *Nucl. Instrum. Methods* **A633** S15–8 https://doi.org/10.1016/j.nima.2010.06.108

[87] Henrich B, Bergamaschi A, Broennimann C, Dinapoli R, Eikenberry E F, Johnson I, Kobas M, Kraft P, Mozzanica A and Schmitt B 2009 PILATUS: A single photon counting pixel detector for x-ray applications *Nucl. Instrum. Methods* **A607** 247–9 https://doi.org/10.1016/j.nima.2009.03.200

[88] Blaj G *et al* 2016 Future of EPix detectors for high repetition rate FELs *AIP Conf. Proc.* **1741** 40012 https://doi.org/10.1063/1.4952884

[89] Ordavo I *et al* 2011 A new PnCCD-based color x-ray camera for fast spatial and energy-resolved measurements *Nucl. Instrum. Methods Phys. Res. Sect. A Accel. Spectrometers, Detect. Assoc. Equip.* **654** 250–7 https://doi.org/10.1016/j.nima.2011.05.080

[90] Strüder L *et al* 1997 A 36 cm^2 large monolythic pn-charge coupled device x-ray detector for the European XMM satellite mission *Rev. Sci. Instrum.* **68** 4271–4 https://doi.org/10.1063/1.1148341

[91] Boyle W S and Smith G E 1970 Charge coupled semiconductor devices *Bell Syst. Tech. J.* **49** 587–93 https://doi.org/10.1002/j.1538-7305.1970.tb01790.x

[92] Fossum E R and Hondongwa D B 2014 A review of the pinned photodiode for CCD and CMOS image sensors *IEEE J. Electron Devices Soc.* **2** 33–43 https://doi.org/10.1109/JEDS.2014.2306412

[93] Matsumoto K, Nakamura T, Yusa A and Nagai S 1985 A new MOS phototransistor operating in a non-destructive readout mode *Jpn. J. Appl. Phys.* **24** L323 https://doi.org/10.1143/JJAP.24.L323

[94] Fowler B, Liu C, Mims S, Balicki J, Li W, Do H, Appelbaum J and Vu P 2010) A 5.5M pixel 100 frames/sec wide dynamic range low noise CMOS image sensor for scientific applications *Proc. SPIE 7536, Sensors, Cameras, and Systems for Industrial/Scientific Applications XI, 753607* https://doi.org/10.1117/12.846975

[95] Griffiths R E, Polucci G, Mak A, Murray S S and Schwartz D A 1981 *Single Photon X-Ray Imaging with Charge-Coupled Devices (CCDs)* (SPIE Proceedings 0290, Solid-State Imagers for Astronomy) pp 62–9 https://doi.org/10.1117/12.965838

[96] Tsunemi H, Mizukata K and Hiramatsu M 1988 Characteristics of optical CCD as an x-ray image sensor *Jpn. J. Appl. Phys.* **27** 670 https://doi.org/10.1143/JJAP.27.670

[97] Tsunemi H, Kawai S and Hayashida K 1991 Performance of the charge-coupled device for direct x-ray detection in the energy range of 1–9 keV at the synchrotron radiation facility *Jpn. J. Appl. Phys.* **30** 1299 https://doi.org/10.1143/JJAP.30.1299

[98] Mathieson K, Passmore M S, Seller P, Prydderch M L, O'Shea V, Bates R L, Smith K M and Rahman M 2002 Charge sharing in silicon pixel detectors *Nucl. Instrum. Methods Phys. Res.* **A487** 113–22 https://doi.org/10.1016/S0168-9002(02)00954-3

[99] Zhao W and Sakurai K 2017 Seeing elements by visible-light digital camera *Sci. Rep.* **7** 45472 https://doi.org/10.1038/srep45472

[100] Zhao W and Sakurai K 2017 CCD camera as feasible large-area-size x-ray detector for x-ray fluorescence spectroscopy and imaging *Rev. Sci. Instrum.* **88** 063703 https://doi.org/10.1063/1.4985149

[101] Zhao W and Sakurai K 2017 Realtime observation of diffusing elements in a chemical garden *ACS Omega* **2** 4363–9 https://doi.org/10.1021/acsomega.7b00930

Chapter 4

Advanced imaging

The previous chapters introduced the importance and applications of projection-type x-ray fluorescence (XRF) imaging. It can reveal the spatial distribution of various chemical elements in inhomogeneous samples. Sometimes, it even captures real-time changes in their spatial distribution. On the other hand, when studying the inhomogeneity of samples, chemical composition is only one aspect of characterization. In many cases, even for the same chemical element, scientists are also interested in the distribution of its different chemical states (valence number, bonding states) or the distribution of different crystalline phases in which it has been involved. If the target element is located on the sample surface or at the interfaces of nanolayers, researchers may also be interested in its nanoscale distribution in the depth direction. The inhomogeneity of these aspects must be characterized using many other imaging techniques, not just XRF imaging. Fortunately, using the same instrument, one can extend projection-type XRF imaging to other advanced x-ray imaging techniques, such as x-ray absorption fine structure (XAFS) imaging, X-ray diffraction (XRD) imaging, x-ray standing-wave imaging, etc. This chapter provides a brief introduction to these advanced imaging techniques.

4.1 X-ray absorption fine structure imaging

In x-ray absorption spectra on the higher-energy side of the absorption edges, some oscillations caused by interference between photoelectron waves are observed. In addition to the chemical shifts of the absorption edge position, some complicated structures are often observed near the absorption edge. Such structures are called XAFS [1–5]. Their analysis is a well-established technique that can provide information about the local atomic structure and chemical state of an element. Experimentally, the sample is illuminated with a highly monochromatic x-ray beam and the intensity of the absorption by the target element is measured. As the x-ray

doi:10.1088/978-0-7503-3215-6ch4

energy is scanned around the absorption edge of the target element, the absorption intensity changes accordingly.

XAFS can be divided into two energy ranges. The first is x-ray absorption near-edge structure (XANES), which appears in a narrow energy range of about 100 eV centered at the absorption edge, where the absorption intensity has a sharp increase. The second, extended x-ray absorption fine structure (EXAFS), appears at about 50 to 1000 eV above the absorption edge, where the absorption intensity has slight periodic oscillations. By locating the precise energy of the absorption edge, observing the shape of the sharp increase, and analyzing the periodicity of the oscillations, it is possible to determine the valence state of the target element and the distribution of atomic spacings between the target element and the surrounding atoms.

To measure XAFS spectra, the most standard, conventional method is the transmission mode. This uses two detectors, one placed in front of and the other behind the sample. In synchrotron facilities, the double-crystal monochromator is rotated to scan the energy of the primary x-rays. At each energy point, both primary and transmitted x-ray intensities are recorded, which can be easily converted into an x-ray absorption spectrum. In the transmission mode, however, the absorption coefficient for each point to be measured has some inherent limitations, such as the perfect uniformity of the sample. Even with a uniform sample, the requirements for sample thickness and concentration are quite strict. If the sample is too thin or if the concentration of the target element in the sample is too low, absorption does not provide contrast. If the sample is too thick, or if the concentration of the target element in the sample is too high, the transmission intensity may be close to zero.

On the other hand, it is possible to measure absorption using an indirect method, such as x-ray fluorescence detection [1–5]. The x-ray fluorescence mode is powerful in studies of the chemical states of trace elements and thin films [6–9]. In addition, the method is easily extended to imaging. This is significant because the analysis of inhomogeneous samples is important in current scientific research. In the case of conventional transmission XAFS, not only are there sample thickness and concentration requirements but also the spatial resolution may be limited. To avoid losing spatial resolution, it is highly recommended to keep the distance between the sample and the 2D detector as small as possible, preferably using a so-called contact imaging geometry. However, this may not be very convenient when measuring real samples.

Figure 4.1 shows an experimental setup for XAFS imaging based on the detection of x-ray fluorescence from realistic inhomogeneous samples [10]. In fact, the instrument layout is essentially the same as that for projection-type XRF imaging, requiring an x-ray optical component inserted between the sample and the detector to project XRF images. The optical component can be a micro pinhole, a collimator plate, or a polycapillary optic. In many scientific cases, the intensity of the fluorescent x-rays is proportional to the intensity of the absorption, i.e. the sample is thin enough and/or the concentration of the elements is dilute enough. Under such conditions, it is necessary to collect the XRF images of the target element near the absorption edge. The 2D x-ray detector used in this approach may or may not have

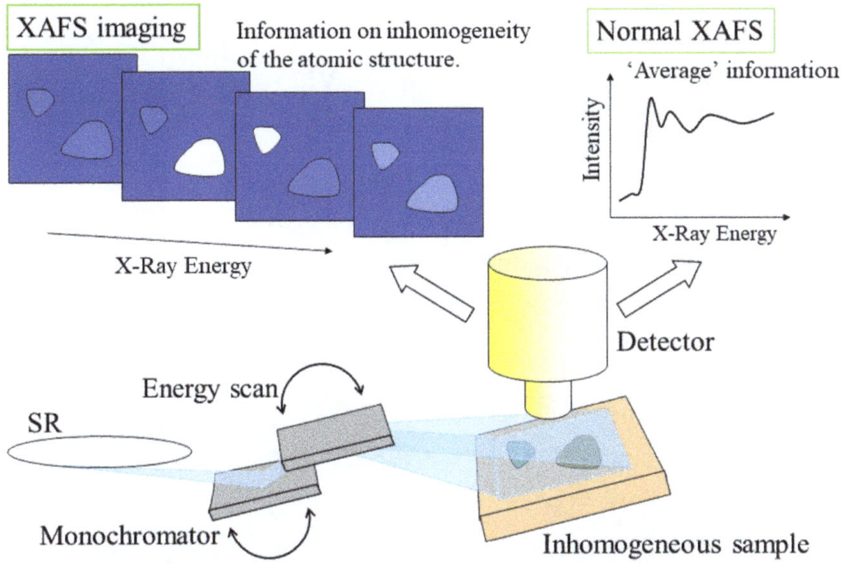

Figure 4.1. A schematic illustration of the experimental geometry and data-collection system used for conventional XAFS and XAFS imaging, both of which are based on x-ray fluorescence detection. XAFS imaging gives XAFS data for each point in the inhomogeneous sample, while the conventional method just averages all points. Reproduced from [10]. Copyright IOP Publishing Ltd. All rights reserved.

energy resolution, but it is preferable to have it. With energy resolution, the XRF emitted by the target element can be separated from the background of elastic and inelastic x-ray scattering as well as XRF from other elements. This allows the target element to be present at low concentrations in the sample without worrying about the effect of a poor signal-to-background ratio on the analysis.

As shown in figure 4.2, a gabbroic rock collected at Mount Tsukuba, Japan, was studied by XAFS imaging [11]. The experiment was performed at BL-16A1, Photon Factory, KEK, Japan, using monochromatic primary x-rays of tunable energy. The detector was a CCD camera (e2V CCD47-10 Class 1, 1024×1024 pixels, pixel size 13 mm × 13 mm, Peltier cooling −30 °C). The optical component was a collimator plate that projected 1:1 XRF images. As shown in figure 4.2(a), the rock sample had two typical mineral textures, black amphibole (area A) and white feldspar (area B). During the experiment, the energy of the primary x-rays was finely scanned around the Fe K-absorption edge (7112 eV) and a series of XRF images were taken, some typical examples of which are shown in figure 4.2(c). The exposure time was ten seconds for each image, and the number of images was typically around 250. For analysis, the integrated intensity in area A and area B were plotted as a function of x-ray energy, as shown in figure 4.2(b). The data were then converted to the form $k\chi(k)$, where k is the wave number of the photoelectron, and Fourier transformed. Although the iron concentrations in both areas were quite different, it was found that the respective local structures were quite similar in areas A and B. The most likely reason is that small yellowish-brown particles mainly contribute to the Fe K-edge XAFS, and they have quite similar atomic structures. The XAFS analysis

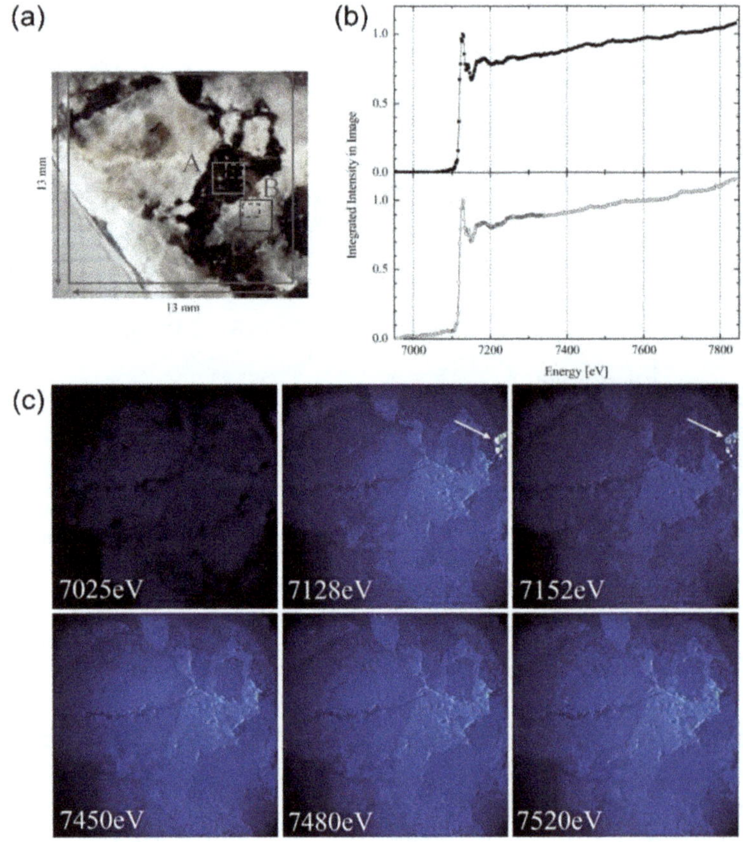

Figure 4.2. (a) A photo of a gabbroic rock. The rectangle of 13 mm × 13 mm encloses the viewing area. (b) XAFS spectra obtained by integrating the intensity in the selected areas in XRF images. The upper spectrum corresponds to area A (black part) and the lower spectrum corresponds to area B (white part). (c) A series of XRF images taken at around the iron K-absorption edge. Adapted from figures 1, 4 and 5 in reference [11]. CC BY 2.0.

shows that the iron in both black amphibole (area A) and white feldspar (area B) is mainly derived from olivine, which is a solid solution of forsterite (Mg_2SiO_4) and fayalite (Fe_2SiO_4), while their ratio is quite different depending on the iron concentration. In addition, the main difference between the two areas is the phase of the oscillations at higher k, which indicates a deviation in the midrange atomic distances.

Figure 4.3 shows an application involving a partially rusted surface of metallic copper [10]. The experiment was performed under similar conditions as in figure 4.2. The x-ray energy was scanned around the Cu K-absorption edge (8979 eV) at steps of 1–2 eV with 1 s of exposure, resulting in a total of 180 images. The integrated intensities for regions 1 (a greenish-blue color) and 2 (metallic copper color) were taken from the collected x-ray images and plotted as a function of x-ray energy. Clearly, the absorption edge for region 1 is shifted due to the chemical change.

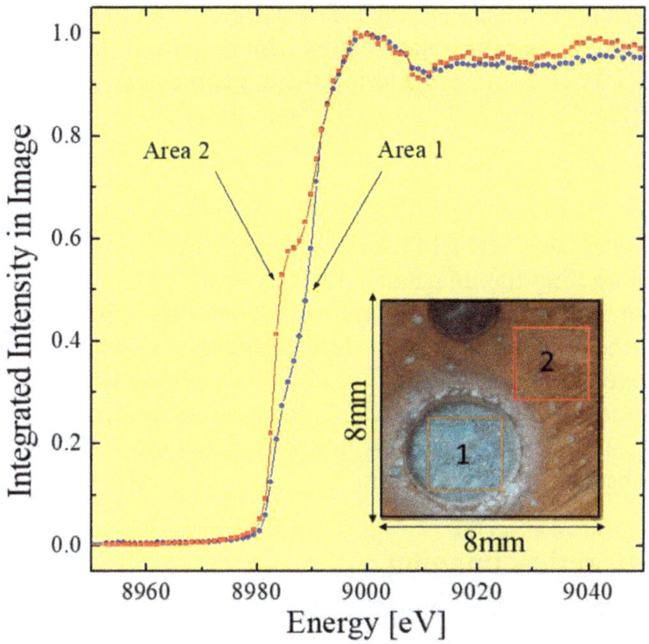

Figure 4.3. XAFS spectra taken from a series of XRF images shown as a function of x-ray energy near the copper absorption edge. Reproduced from [10]. Copyright IOP Publishing Ltd. All rights reserved. The analysis areas were a greenish-blue area (orange box, area 1) and the copper area (red box, area 2) on a copper plate, with a field of view of 8 mm × 8 mm. The results show that the rusted area 1 exhibits a chemical shift corresponding to oxidation.

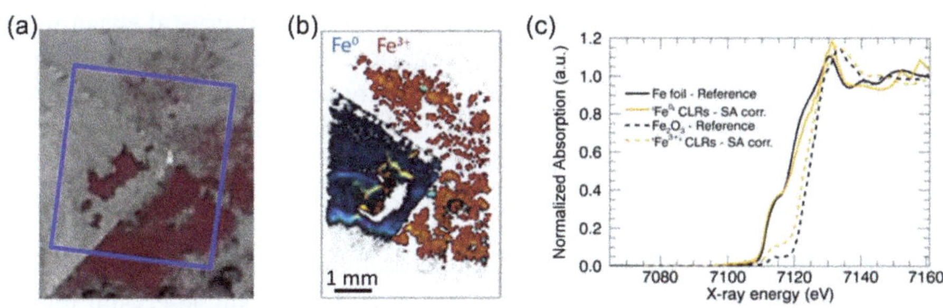

Figure 4.4. (a) A photo of an iron and iron oxide sample composed of a 4 μm thick iron foil, as well as FeO and Fe_2O_3 particles. The rectangle encloses the 3 mm × 3 mm viewing area. (b) The results, showing the Fe^0 and Fe^{3+} distributions. (c) Fluorescence-mode XANES spectra in the regions of iron foil and Fe_2O_3 following self-absorption (SA) correction. Modified with permission from figures 5 and 7 in reference [12], copyright (2014) American Chemical Society.

Another example is a study of a sample composed of a 4 μm thick iron foil as well as ferrous oxide and ferric oxide particles [12]. A photo of the sample is shown in figure 4.4(a). The experiment was carried out at BM26A, the x-ray absorption spectroscopy station of the Dutch–Belgian beamlines (DUBBLE) at the European

Synchrotron Radiation Facility (ESRF). The detector utilized was a pnCCD detector with energy resolution, and the optical component employed was a polycapillary optic which projected 1:1 XRF images. During the experiment, a series of iron Kα XRF images were acquired along with the energy scan. In addition, the transmission-mode XANES spectra were measured for the iron foil and the ferric oxide particles, respectively, which are plotted as references in figure 4.4(c). It was observed that the greatest difference between these two spectra occurred at 7.120 keV, whereas the minimum difference was at 7.143 keV. Consequently, a differential image was generated by subtracting the image at 7.120 keV from the image at 7.143 keV. As a result, Fe^{3+} contributes the most prominent signals in the differential image. The results shown in the differential image, which demonstrate the distributions of Fe^0 and Fe^{3+}, are displayed in figure 4.4(b). It exhibits a clear agreement with the sample photo. Furthermore, the fluorescence-mode XANES spectra integrated from the iron Kα XRF images are plotted in figure 4.4(c), which also shows a clear agreement with the reference transmission-mode spectra.

4.2 X-ray diffraction imaging

XRD happens when the wavelength of incident x-rays λ, the angle between the incident x-rays and a specific lattice plane θ_B, and the interplanar spacing of the lattice plane d satisfy Bragg's law:

$$2d \sin \theta_B = n\lambda. \tag{4.1}$$

Here, n can be any positive integer, but the diffraction intensity is the strongest when n equals one. The diffracted x-rays emitted at the same angle θ_B appear as if x-rays are reflected by the lattice plane. In a crystal structure, there are multiple sets of lattice planes with different interplanar spacings d. Different crystal structures may have characteristic combinations of interplanar spacings $\{d_1, d_2, d_3, \ldots\}$, which can be sequentially measured by XRD. When these are combined with the diffraction intensity, the crystal structure can be identified. This is the theoretical basis for identifying crystal structures through XRD [13–15].

X-ray diffraction was discovered by Max von Laue in 1912, over 110 years ago. This technique has been extensively developed and applied for studying the structures of inorganic and organic crystals, including the crystals of biological macromolecules such as nucleic acids and proteins. Unlike single-crystal materials, XRD imaging for polycrystalline inhomogeneous samples has been less well known despite its scientific significance and the demand for it. In 1982, Chikaura, Yoneda, and Hildebrandt attempted to see the inhomogeneity of polycrystalline materials using a scanning scheme [16]. Since 1995, Wroblewski has reported his series of pioneering studies of projection-type XRD imaging for polycrystalline materials [17–19]. XRD imaging can reveal the distributions of different crystalline phases and strains in inhomogeneous polycrystalline samples [20–29].

The authors have found another promising projection-type x-ray diffraction imaging technique for polycrystalline materials. As shown in figure 4.5, the method uses grazing incidence geometry and the 2D detector. Note that a collimator plate is placed close to the detector surface, similar to Wroblewski's pioneering work. If the x-ray source has

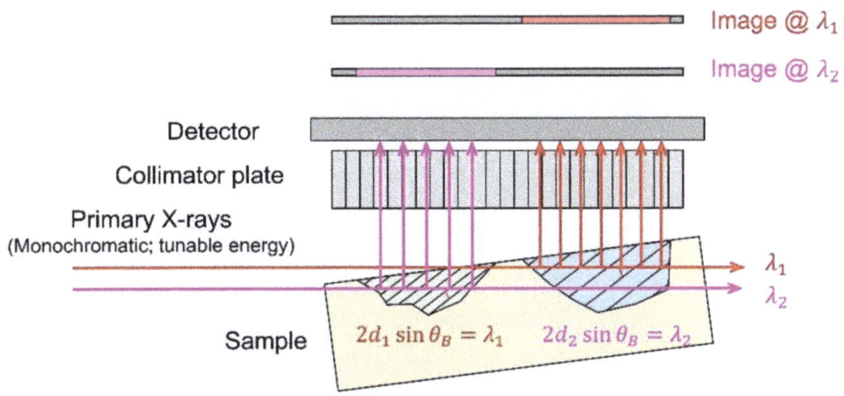

Figure 4.5. A schematic illustration of projection-type x-ray diffraction imaging for polycrystalline materials using tunable monochromatic x-rays.

continuum spectra and the tunable monochromatic x-ray photons are available, the imaging can be done with fixed geometry. During the experiment, the upstream monochromator is rotated to change and tune the energy, i.e. the wavelength, of the incident x-rays. In terms of instrumentation, this type of XRD imaging is very similar to the XAFS imaging described in figure 4.1. The main difference is in the choice of x-ray energy. The energy must be a Bragg energy that satisfies the Bragg condition (equation (4.1)) instead of the absorption edge energy for XAFS imaging. The 2D detector is intended to receive diffracted x-rays and does not require energy resolution. In this layout, the selected Bragg angle θ_B is fixed by the included angle between the incident x-ray beam and the collimator axis. The included angle is preferably $\pi/2$, but it can also be slightly adjusted depending on specific requirements.

For a specific crystal structure, x-ray diffraction happens when the wavelength of the incident x-rays is adjusted to match its one characteristic interplanar spacing. In the resulting image captured by the detector, a high-intensity region appears where the crystal structure is distributed. By capturing images at several specific x-ray wavelengths, the distributions of different crystal structures can be revealed.

It is important to note that this experimental approach enables imaging to be performed even for polycrystalline and/or powder samples in which the lattice planes are oriented arbitrarily. For a single crystal, imaging known as x-ray topography has a long history, and a specific measurement system exists [30].

Figure 4.6 shows an XRD image of a patterned sample made of metallic chromium thin film [24]. The authors use such samples to check the spatial resolution of XRF imaging. In this case, the contrast of the image is not due to x-ray fluorescence, but by x-ray diffraction. The bright part corresponds to the 110 diffraction from the chromium (110) crystal plane. As the incident x-ray energy is 5.753 keV, which is lower than the Cr K-absorption edge (5.989 keV), x-ray fluorescence is not generated by the patterned metallic film. However, x-ray diffraction takes place if the angle and the x-ray energy satisfy the Bragg condition. In this case, the angular position of the camera (2θ) was set to 64°.

Since the early days of modern materials chemistry, even as early as the 1950s, XRD has been used to perform the quantitative analysis of mixtures of anatase and

Figure 4.6. An x-ray image of a patterned sample. The contrast of the image is caused by x-ray diffraction from the chromium (110) crystal plane. The x-ray energy used is 5.753 keV, which is slightly lower than the Cr–K absorption edge (5.989 keV). The image was not obtained by X-ray fluorescence of chromium, but by X-ray diffraction. [24].

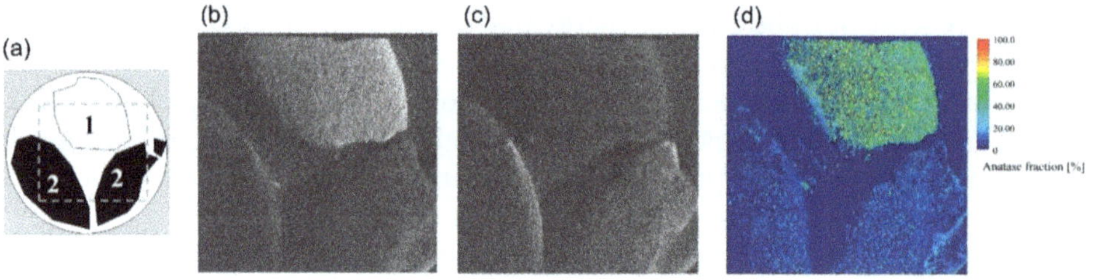

Figure 4.7. XRD imaging of an anatase–rutile mixture. (a) A schematic drawing of the location of anatase-rich (1) and rutile-rich (2) regions. (b) and (c) X-ray diffraction images caused by anatase 200 reflection and rutile 210 reflection, respectively. They are from anatase (100) and rutile (210) crystal planes, respectively. (d) The results for the anatase fraction as a weight percentage. Adapted with permission from figures 1, 2, and 4 in reference [25], copyright (2014) American Chemical Society.

rutile, both of which are well-known stable phases of titanium dioxide (TiO_2) [31]. In real-world applications, TiO_2 is most likely to be a mixture of these rather than either pure anatase or pure rutile [32]. Furthermore, control of the transformation between anatase and rutile is one of the most important issues in synthesis. Typically, ordinary XRD techniques need uniform samples for measurement and analysis. In the case of a mixture of several materials, it is usually assumed that the concentration (or the weight) ratio is almost the same at all points in the mixture sample. However, as is often the case for thin films and powders, real samples are not always uniform. One needs information in the form of a map rather than a single value of the ratio in such cases. The XRD imaging discussed here can satisfy this demand.

Figure 4.7 shows an XRD image of a pellet made from a mixture of rutile and anatase [25]. In this sample, anatase-rich mixtures and rutile-rich mixtures are located in different regions. While it is impossible to distinguish them by XRF imaging, XRD imaging works well. This experiment was carried out at Beamline 16A1, Photon Factory, KEK, Japan. The detector utilized was a CCD camera (Texas Instruments TC281, exposure area 1024×1024 pixels, pixel size 8 μm squared, Peltier cooling −30 °C) equipped with an internal collimator plate. The x-ray images were acquired using 4.9 keV primary x-rays at angles of 84° and 76° between the primary x-rays and the collimator axis, respectively. They correspond to the anatase 200 and rutile 210 reflections, respectively. They are from the anatase (100) and rutile (210) crystal planes, respectively. Though it is quite easy to see the differences between the two images, much more quantitative analysis can be a great help. This technique can give the spatial distribution of the fraction ratio of the anatase (= (anatase weight)/(anatase weight + rutile weight)) for the anatase–rutile mixture, a quite typical form of real-world TiO_2. To obtain this distribution, the intensity ratio of the anatase 200 reflection and the rutile 210 reflection is analyzed for each point. Finally, the above two images are processed to form the anatase fraction map. Using projection-type imaging, it has become possible to visualize anatase-rich and rutile-rich regions very quickly by simply choosing two different diffraction angles. The mapping of the anatase fraction factor is useful for quantitative discussion. As this technique can give a 2D image of x-ray diffraction intensity from an inhomogeneous sample, many other similar important applications involving inhomogeneous complex materials can be considered in addition to the anatase–rutile case. Furthermore, as the method can be extended to real-time movies, freezing and melting processes in liquid metals can be observed as x-ray diffraction images during changes in temperature [26].

One of the other interesting applications of XRD imaging is to see the distribution of lattice distortion in steel materials of different hardness [27, 28]. In this study, Vickers test blocks of high-carbon steel with different hardness were examined. This imaging method can visualize the change in *d*-value in the vicinity of the Vickers mark. The strength of this technique lies in its ability to perform quantitative evaluation. Grain size and hardness are correlated. Generally, steel with a large grain size is soft, while that with smaller grain size is hard. Interestingly, the shape and size of the crystal grains were directly observed with a lateral spatial resolution of 12–18 μm, which depends on the aperture of the collimator plate and the distance between the sample and the detector pixels. In this study, as shown in figure 4.8, round grains 40–50 μm in diameter were observed for the lower-hardness sample. On the other hand, in the case of the higher-hardness sample, no good images were obtained because the grain size was less than the lateral spatial resolution. However, the image obtained visualizes the strains introduced by the Vickers indentation. Some shifts in the XRD peak position were observed in the vicinity of the Vickers mark (<1 mm).

4.3 Integrating x-ray imaging techniques

As already discussed, the techniques used to image chemical composition (XRF), local structure and chemical states (XAFS), and crystalline structures (XRD) use

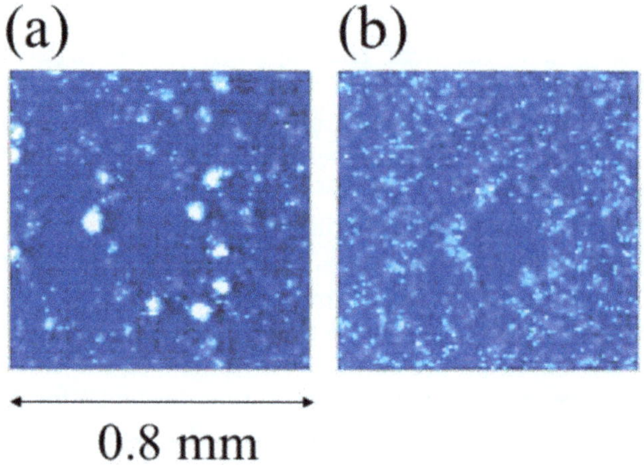

Figure 4.8. XRD images of Vickers test blocks of high-carbon steel with different hardness. Reproduced from [27]. X-ray spots due to 200 reflection were observed at $2\theta = 90°$. To avoid the influence of Fe K x-ray fluorescence, the energy was set lower than the absorption edge. (a) A lower-hardness sample with larger grain. (b) A higher-hardness sample with smaller grain. Both samples have an indentation mark at the center.

nearly the same instrument setups, the only difference being the energy of the primary x-rays. XRF imaging requires the energy to be higher than the absorption edge of the target element; XAFS imaging involves continuously scanning the energy around the absorption edge; whereas XRD imaging requires selecting appropriate energies, preferably energies lower than the absorption edges of the major elements in the sample, to avoid the background of XRF. For this reason, researchers can perform all three imaging techniques in the same experiment by adjusting the energy of the primary x-rays. This combinational imaging enables a comprehensive characterization of the sample's inhomogeneity of elements, chemical states, and crystal structures. It can serve as a powerful tool for studying complex natural samples such as mineral ores and industrial products like corroded steel alloys.

Figures 4.9 and 4.10 show an example of integrated x-ray imaging [29]. The sample was an artificial mixture of hematite (α-Fe_2O_3), magnetite (Fe_3O_4), and lepidocrocite (γ-FeOOH) powders. Its photo is displayed in figure 4.9(c). The experiment was carried out at Beamline NW2, Photon Factory Advanced Ring, KEK, Japan. The detector utilized was a CCD camera equipped with a collimator plate at its front. The CCD camera had no inherent energy resolution. During the experiment, 200 XRF images were acquired while scanning the x-ray energy around the iron K-absorption edge using an energy interval of 0.5 eV. Two XRF images acquired at the edge, namely at 7.118 keV and 7.130 keV, are displayed in figures 4.9(a) and (b), respectively. Although the images have no energy resolution, it can be confirmed that the majority of the signals came from iron. The integrated XAFS spectra within different regions corresponding to Fe_2O_3, Fe_3O_4, and γ-FeOOH are plotted in figure 4.9(e), where the differences between these spectra are easily observable.

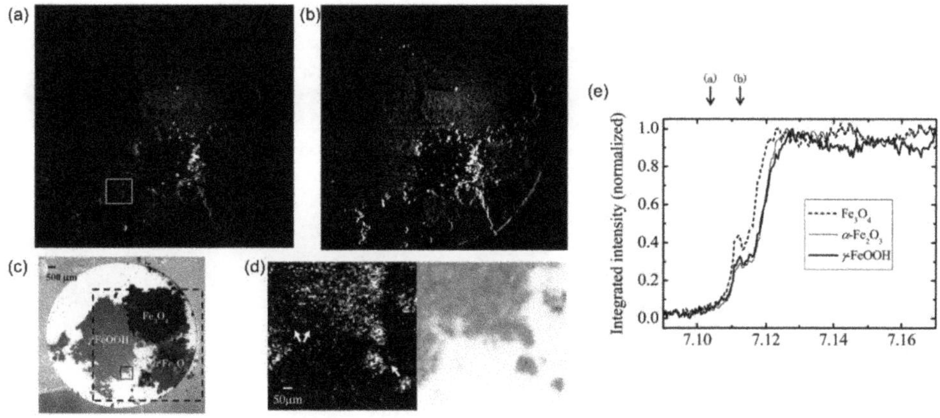

Figure 4.9. (a) and (b) XRF images of the iron oxide sample acquired at 7.118 keV and 7.130 keV, respectively. The area enclosed in the square in (a) is enlarged in (d). (c) A photo of the sample, which is a mixture of α-Fe_2O_3, Fe_3O_4, and γ-FeOOH. The dashed rectangle indicates the 8 mm × 8 mm viewing area. (e) Fluorescence-mode XAFS spectra of the three iron oxides. The labeled arrows indicate the x-ray energies used to acquire (a) and (b). Adapted from figures 1 and 2 in reference [29] with permission from the Royal Society of Chemistry.

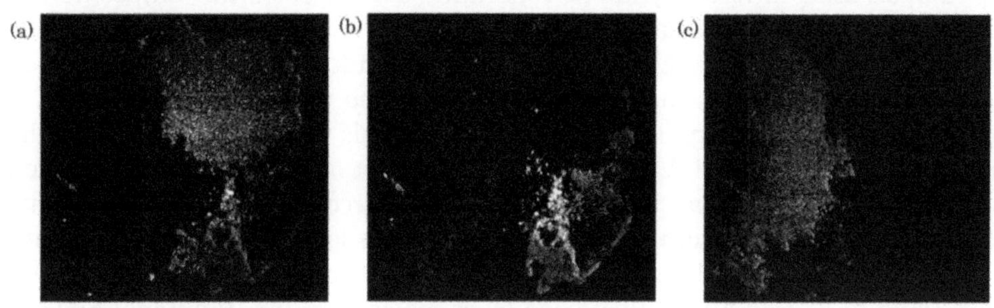

Figure 4.10. XRD images of the iron oxide sample. (a) A 440 reflection of Fe_3O_4 and a 214 reflection of α-Fe_2O_3 obtained using 5.937 keV x-rays. (b) A 300 reflection of α-Fe_2O_3 and a weak 180 reflection of γ-FeOOH obtained using 6.056 keV x-rays. (c) A 231 reflection of γ-FeOOH using 5.777 keV x-rays. Adapted from figures 3 and 5 in reference [29] with permission from the Royal Society of Chemistry.

For XRD imaging, the included angle between the primary x-rays and the collimator axis remained at 90°, i.e. the diffraction angle θ_B was selected to be 45°. The x-ray energies selected were 5.937 keV, 6.056 keV, and 5.777 keV, which produced the XRD images shown in figures 4.10(a), (b), and (c), respectively. Since these selected x-ray energies are below the iron K-absorption edge, the XRD images are free from the background of iron K-fluorescence x-rays. Information regarding the corresponding lattice plane reflections is provided in the figure caption. It is easy to confirm the agreement between the XRD images and the sample photo.

4.4 The application of x-ray color imaging in combinatorial imaging

Scientists have struggled to find a good strategy with which to discover attractive and promising functional materials. One of the methods established in the 20th

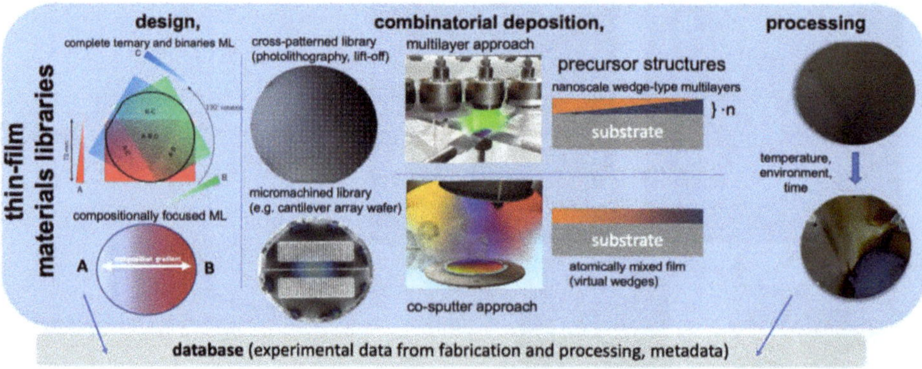

Figure 4.11. The concept of combinatorial materials synthesis (for the case of thin films). The basic idea is to fabricate a sample corresponding to binary or ternary graded mixtures on a single substrate called a combinatorial library. The method is suitable for efficient, high-throughput surveying, and the data are added to a database. The illustration is taken from figure 1 in reference. Reproduced from [34]. CC BY 4.0.

century is combinatorial materials synthesis [33, 34]. The idea is to prepare many tiny samples using different synthetic conditions, such as chemical composition, pH, temperature, etc. All the samples are prepared on a single substrate. The substrate with many tiny samples is called a combinatorial library. Figure 4.11 schematically shows the strategy used in the design, fabrication, and processing of thin films. The combinatorial library is spatially graded. If the goal is to find the optimal synthesis conditions for a bright UV-sensitive scintillator, it is efficient to study such a combinatorial library using an ultraviolet light source. In the early 21st century, because of the thickness limit of the insulating oxide layer in conventional metal--oxide–semiconductor field-effect transistors, scientists in the semiconductor industry were trying to find an alternative oxide-layer material to the traditional SiO_2. The requirement was that it must have a high dielectric constant to allow it to be used in MOS structures even in the case of nanoscale design, i.e. close to the limit of Moore's law. Such high-κ materials are based on HfO_2, but the question is how to optimize the chemical composition when other components are included. The combinatorial approach was used in the search [35, 36]. In addition to synthesis and functional evaluation, the characterization of a combinatorial library is also important. Therefore, it is important to establish a novel characterization method.

The main idea is that today's x-ray color imaging can be used as a technique for combinatorial imaging [24, 37–39]. Figure 4.12 shows an example of combinatorial imaging for high-κ materials [24]. The sample was prepared by the pulsed-laser deposition method, and each part of the substrate has a different chemical composition. The technique can quickly visualize where crystalline HfO_2, Al_2O_3, and Y_2O_3 exist. As scientists were more interested in the noncrystalline regions, this imaging was able to work as an efficient screening scheme.

Figure 4.13 shows the use of the manganese cobalt oxide nanoparticles in a combinatorial study [37]. The target material was synthesized by evaporation drying and the treatment of a mixed aqueous solution of manganese and cobalt salts at

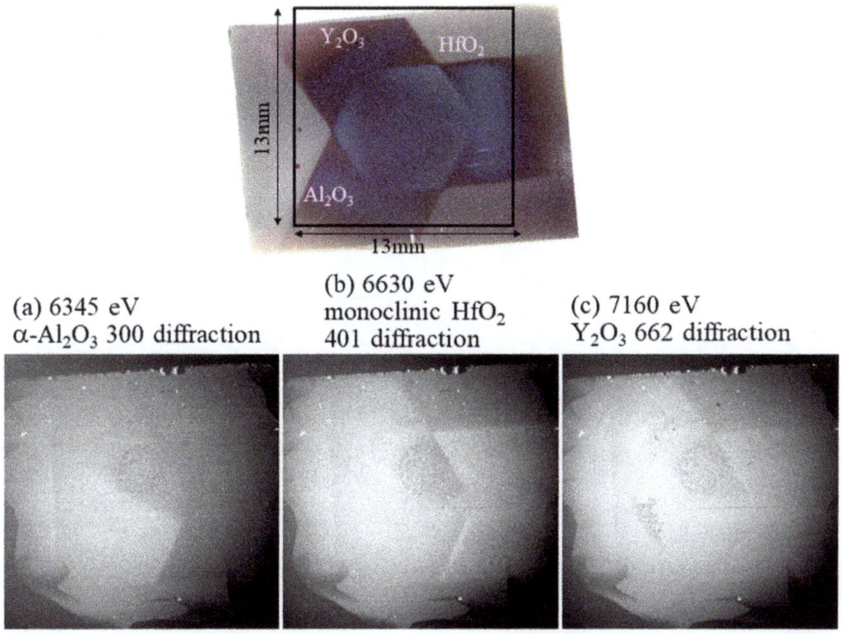

Figure 4.12. XRD imaging of a combinatorial library for high-κ thin films which was prepared using a pulsed-laser deposition technique. Three targets are included, namely Al_2O_3, Y_2O_3, and HfO_2 [24]. The chemical composition of each point was graded, and some regions were found to be in a noncrystalline phase. Sample courtesy to Dr T Chikyow, National Institute for Materials Science. The details of these high-κ thin films system have been described elsewhere [36].

temperatures below 400 °C. To optimize the synthetic conditions, such as a reduction in the processing temperature to control the size of nanoparticles or the addition of nitric acid to control the defect structure, microsamples with different processing temperatures and nitric acid concentrations were arranged on the same alumina substrate. First, their chemical composition was studied by acquiring x-ray fluorescence images. XAFS imaging was then performed by scanning the incident x-ray energy near the K-edges of manganese and cobalt. The method is equivalent to the simultaneous measurement of a large number of XAFS spectra, as there are approximately 1000×1000 pixels. In this example, the samples were arranged in a 3 × 3 matrix, so the XAFS spectra of all nine samples could be obtained simultaneously. As shown in the figure, there was a difference in the chemical shift of the manganese K-edge, indicating that at low temperatures and low amounts of nitric acid, thermal decomposition of the salt was insufficient, resulting in a lower oxidation number than the target spinel-type oxide.

4.5 Other advanced x-ray imaging

As already described, XAFS and XRD imaging can be performed by scanning the primary x-ray energy. Many images are collected at each x-ray energy during the monochromator scan. The data collection is very similar to that of movie experiments,

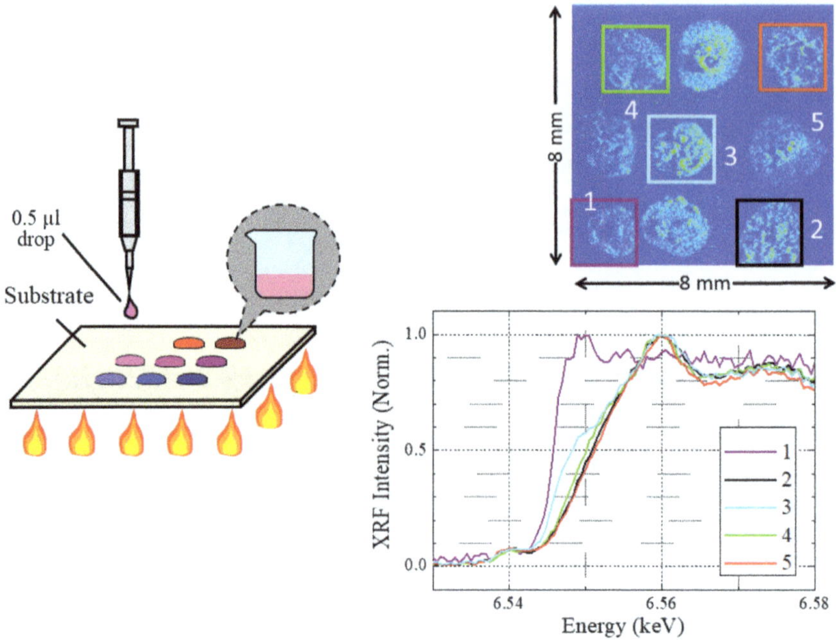

Figure 4.13. The combinatorial imaging of manganese cobalt oxide nanoparticles. Reproduced with permission from [37]. Copyright 2005 The Chemical Society of Japan. Many samples with different synthesis conditions can be arrayed on the same substrate to form a combinatorial library. In this case, nine samples were simultaneously analyzed by obtaining XAFS images (1000×1000 pixels, 1 s per image) by scanning the incident x-ray energy near the absorption edges. The total measurement time was 13 min. The XAFS spectra were plotted by extracting intensity data from the images.

which use sequential image capture. It is also possible to extend the same approach to other advanced x-ray imaging. For example, XRF images can be collected as a function of the angle of incidence, which can be changed in small and precise steps.

One such attempt is the application of sequential image capture in x-ray standing-wave (XSW) imaging. In general, XSWs are seen in the Bragg reflection of perfect single crystals and some periodic multilayers [40]. As shown in figure 4.14, the XSW field is created by interference between incident x-rays and reflected x-rays, which provides an atomic-scale metric for measuring the precise position of impurities in their host materials. In 1964, the phenomena and technique were first discovered for germanium single crystals by Batterman [41, 42]. The essential point is that the x-ray fluorescence from atoms in the crystal exhibits a sharp characteristic angular dependence near the Bragg condition. Theoretically, this phenomenon is explained by a dynamical effect that only occurs in a perfect single crystal [43, 44]. Since the x-ray fluorescence comes from a specific chemical element in the crystal, the technique can be used to determine the exact position of the impurity atoms in the lattice plane. During angular scanning near the Bragg angle, the XSW shifts from one crystal plane to the neighboring plane. By analyzing the x-ray fluorescence intensity profile, the position, i.e. the distance from the crystal plane, can be

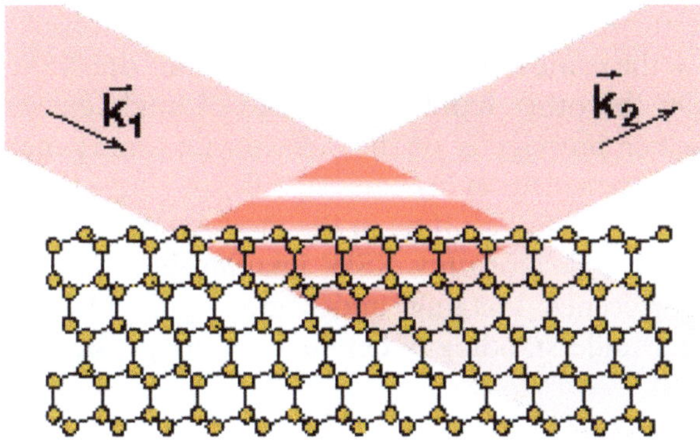

Figure 4.14. A schematic illustration of an x-ray standing wave generated by interference between incident and specularly reflected plane waves. [40, 45]. When scanning the angle in the vicinity of the Bragg condition, the standing wave shifts from one plane to the neighboring plane.

determined. Subsequently, the technique was extended to the study of adsorbates on a crystal surface, i.e. not only inside bulk crystals. Furthermore, the technique has been extended to periodic multilayer thin films and total reflection mirrors [45–47]. In this case, the angle becomes much smaller than that of ordinary Bragg diffraction in the bulk crystal, with the result that an XSW with a fairly large periodic length forms on the surface. This has opened up the possibility of seeing molecules of various sizes. In the case of surface analysis, the detection of photoelectrons and Auger electrons instead of x-ray fluorescence is promising.

The initial idea of XSW imaging may be quite close to that of x-ray fluorescence holography [48, 49], which uses interference between x-ray fluorescence in a single crystal. M J Bedzyk, who is a pioneer of the XSW technique for multilayer Bragg reflection and general total reflection cases, has already reported 3D mapping of an impurity by XSW imaging [50, 51]. The importance of such visualization is now widely accepted [52]. On the other hand, the inhomogeneous distribution of impurity elements throughout the sample is significant. In conventional XSW analysis, the scanning area is determined by the area illuminated by the incident x-rays, which can be as large as a few square millimeters or even larger. In addition to determining the exact height or depth position normal to the surface, it is important to know how the impurities are distributed over such a large area. Therefore, the application of projection x-ray fluorescence imaging in macroscopic XSW imaging is promising. In this case, the only requirement is to replace the x-ray detector used for x-ray fluorescence with the system described in this book. The experiment is essentially the same as for projection x-ray fluorescence imaging. The specific point for XSW imaging is that one needs to precisely scan the angle of incidence and record a series of XRF images. Since the area of nanolayers and interfaces is often a few square centimeters to square millimeters, an imaging resolution in the submillimeter to micrometer range is already sufficient to match the macroscopic in-plane inhomogeneity of the entire sample. In 2019, iron impurities in a nickel (1.4 nm)/carbon (3.6 nm) periodic multilayer were studied using macroscopic XSW imaging [53].

The idea can be further applied to any kind of angle-resolved x-ray fluorescence experiments. Below the critical angle of total reflection, the XSW is formed on the surface [40, 45]. On the other hand, the change of angle causes a change in the evanescent wave, which propagates on the surface with some penetration. In the case of thin films, when the penetration becomes significant above the critical angle, some interference effects take place, resulting in the modulation of the angular profiles of x-ray fluorescence [54, 55]. Not only XSW imaging but also more general angle-resolved x-ray fluorescence imaging under grazing incidence conditions could contribute to a better understanding of detailed impurity distributions at the surface and interfaces.

References

[1] Koningsberger D C and Prins R (ed) *X-Ray Absorption: Principles, Applications, Techniques of EXAFS, SEXAFS and XANES* (New York: Wiley) 1988 https://www.wiley.com/en-cn/exportProduct/pdf/9780471875475

[2] Bunker G 2010 *Introduction to XAFS* (Cambridge: Cambridge University Press) https://doi.org/10.1017/CBO9780511809194

[3] Calvin S 2013 *XAFS for Everyone* (Boca Raton, FL: CRC Press) https://doi.org/10.1201/b14843

[4] van Bokhoven J A and Lamberti C 2016 *X-Ray Absorption and X-Ray Emission Spectroscopy: Theory and Applications* (New York: Wiley) https://doi.org/10.1002/9781118844243

[5] Evans J 2017 *X-Ray Absorption Spectroscopy for the Chemical and Materials Sciences* (New York: Wiley) https://doi.org/10.1002/9781118676165

[6] Sakurai K, Iida A and Gohshi Y 1987 Characterization of Co–O thin films by x-ray fluorescence using chemical shifts of absorption edges *Jpn. J. Appl. Phys.* **26** 1937–8 https://doi.org/10.1143/JJAP.26.1937

[7] Sakurai K, Iida A and Gohshi Y 1988 Chemical state analysis by x-ray fluorescence using shifts of iron K absorption edge *Anal. Sci.* **4** 37–42 https://doi.org/10.2116/analsci.4.37

[8] Sakurai K, Iida A, Takahashi M and Gohshi Y 1988 Chemical state mapping by x-ray fluorescence using absorption edge shifts *Jpn. J. Appl. Phys.* **27** L1768–71 https://doi.org/10.1143/JJAP.27.L1768

[9] Sakurai K, Iida A and Gohshi Y 1989 Chemical state analysis by x-ray fluorescence using absorption edges shifts *Adv. X-ray Anal.* **32** 167–76 https://doi.org/10.1154/S0376030800020437

[10] Sakurai K and Mizusawa M 2004 Quick atomic-scale structure imaging by synchrotron x-rays: a new tool for probing realistic inhomogeneous systems *Nanotechnology* **15** S428 https://doi.org/10.1088/0957-4484/15/6/021

[11] Mizusawa M and Sakurai K 2004 XAFS imaging of Tsukuba gabbroic rocks: area analysis of chemical composition and local structure *J. Synchrotron Radiat.* **11** 209–13 https://doi.org/10.1107/S0909049503028024

[12] Bras W, Laforce B, Bauters S, Garrevoet J, Vincze L, Banerjee D, Van Ranst E, Longo A, Vekemans B and Tack P 2014 Full-field fluorescence mode micro-XANES imaging using a unique energy dispersive CCD detector *Anal. Chem.* **86** 8791–7 https://doi.org/10.1021/ac502016b

[13] Guinier A 1963 *X-Ray Diffraction: In Crystals, Imperfect Crystals and Amorphous Bodies* (San Francisco, CA: W.H. Freeman)

[14] Warren B E 1969 *X-Ray Diffraction* Addison-Wesley Series in Metallurgy and Materials (Reading, MA: Addison-Wesley)

[15] Cullity B D 1956 *Elements of X-ray Diffraction* (Reading, MA: Addison-Wesley)

[16] Chikaura Y, Yoneda Y and Hildebrandt G 1982 Polycrystal scattering topography *J. Appl. Cryst.* **15** 48 https://doi.org/10.1107/S0021889882011339

[17] Wroblewski T, Geier S, Hessmer R, Schreck M and Rauschenbach B 1995 X-ray imaging of polycrystalline materials *Rev. Sci. Instrum.* **66** 3560–2 https://doi.org/10.1063/1.1145469

[18] Wroblewski T 2002 MAXIM—a novel method for the x-ray imaging of strain *Mater. Sci. Forum* **404–407** 121–6 https://doi.org/10.4028/www.scientific.net/MSF.404-407.121

[19] Wroblewski T and Bjeoumikhov A 2005 X-ray diffraction imaging of bulk polycrystalline materials *Nucl. Instrum. Methods Phys. Res.* A **538** 771–7 https://doi.org/10.1016/j.nima.2004.09.022

[20] Japan Patent No. 3834652 2006 Sakurai M X-ray diffraction microscope apparatus and x-ray diffraction measuring method by the same (patent application 2003-318922) https://patentscope2.wipo.int/search/en/detail.jsf?docId=JP270215955

[21] Japan Patent No. 4581126 2010 Sakurai M X-ray diffraction analyzer and x-ray diffraction analyzing method (patent application 2005-066097). https://patentscope2.wipo.int/search/en/detail.jsf?docId=JP270797683

[22] Japan Patent No. 4604242 2010 Sakurai M X-ray diffraction analyzer and x-ray diffraction analyzing method' (patent application 2005-066120). https://patentscope2.wipo.int/search/en/detail.jsf?docId=JP270797704

[23] Japan Patent No. 4674352 2011 Sakurai M Titanium oxide analyzing method and titanium oxide analyzer carrying out it (patent application 2005-114013). https://patentscope2.wipo.int/search/en/detail.jsf?docId=JP270840817

[24] Mizusawa M and Sakurai K unpublished data

[25] Sakurai K and Mizusawa M 2010 X-ray diffraction imaging of anatase and rutile *Anal. Chem.* **82** 3519–22 https://doi.org/10.1021/ac9024126

[26] Mizusawa M and Sakurai K 2010 In-situ observation of inhomogeneous texture during melting and freezing of liquid gallium by projection-type x-ray diffraction imaging *Bunseki Kagaku* **59** 499–511 (in Japanese) https://doi.org/10.2116/bunsekikagaku.59.499

[27] Mizusawa M and Sakurai K 2021 2D real space visualization of *d*-values in polycrystalline bulk materials of different hardness *J. Appl. Crystallogr.* **54** 597–603 https://doi.org/10.1107/S1600576721001631

[28] Mizusawa M and Sakurai K 2021 Projection-type x-ray diffraction imaging for polycrystalline materials: application to vickers hardness test blocks *Adv. X-Ray Anal.: Proceedings of the Denver X-ray Conference* **64** 51–8 https://icdd.com/axa-login/?article=56817116519d7b

[29] Eba H, Ooyama H and Sakurai K 2016 Combination of projection-based XRF, XAFS and XRD imagings for rapid spatial distribution analysis of a heterogeneous material *J. Anal. At. Spectrom.* **31** 1105–11 https://doi.org/10.1039/C6JA00024J

[30] Tanner B K and Bowen D K 1992 Synchrotron x-radiation topography *Mater. Sci. Rep.* **8** 371–407 https://doi.org/10.1016/0920-2307(92)90002-I

[31] Spurr R A and Myers H 1957 Quantitative analysis of anatase–rutile mixtures with an x-ray diffractometer *Anal. Chem.* **29** 760–2 https://doi.org/10.1021/ac60125a006

[32] Bakardjieva S, Šubrt J, Štengl V, Dianez M J and Sayagues M J 2005 Photoactivity of anatase–rutile TiO_2 nanocrystalline mixtures obtained by heat treatment of homogeneously precipitated Anatase *Appl. Catalysis* B **58** 193–202 https://doi.org/10.1016/j.apcatb.2004.06.019

[33] 2003 *Combinatorial Materials Synthesis* ed X-D Xiang and I Takeuchi (Boca Raton, FL: CRC Press) https://doi.org/10.1201/9780203912737

[34] Alfred L 2019 Discovery of new materials using combinatorial synthesis and high-throughput characterization of thin-film materials libraries combined with computational methods *NPJ Comput. Mater.* **5** 70 https://doi.org/10.1038/s41524-019-0205-0

[35] Chikyow T, Ahmet P, Nakajima K, Okazaki N, Hasegawa K, Hasegawa T and Koinuma H 2003 Critical issues of new gate insulator/Si interfaces in the future ULSI *Trans. Mater. Res. Soc. Jpn* **28** 15 https://mrs-j.org/pub/tmrsj/vol28_special/vol28_special_015.pdf

[36] Ahmet P, Yoo Y-Z, Hasegawa K, Koinuma H and Chikyow T 2004 Fabrication of three-component composition spread thin film with controlled composition and thickness *Appl. Phys.* A **79** 837–9 https://doi.org/10.1007/s00339-004-2627-9

[37] Eba H and Sakurai K 2005 Combinatorial fluorescence XAFS imaging of manganese complex oxides *Chem. Lett.* **34** 872–3 https://doi.org/10.1246/cl.2005.872

[38] Eba H and Sakurai K 2005 Enhancement of CO_2 absorbance for lithium ferrite—combinatorial application of x-ray absorption fine structure imaging *Mater. Trans* **46** 665–8 https://doi.org/10.2320/matertrans.46.665

[39] Eba H and Sakurai K 2006 Rapid combinatorial screening by synchrotron x-ray imaging *Appl. Surf. Sci.* **252** 2608–14 https://doi.org/10.1016/j.apsusc.2005.07.077

[40] Zegenhagen J and Kazimirov A 2013 *The X-Ray Standing Wave Technique: Principles and Applications* ed J Zegenhagen and A Kazimirov (Hackensack, NJ: World Scientific)

[41] Batterman B W 1964 Effect of dynamical diffraction in x-ray fluoresence scattering *Phys. Rev.* **133** A759–64 https://doi.org/10.1103/PhysRev.133.A759

[42] Batterman B W 1969 Detection of foreign atom by their x-ray fluorescence scattering *Phys. Rev. Lett.* **22** 703 https://doi.org/10.1103/PhysRevLett.3.32

[43] Cowan P L, Golovchenko J A and Robbins M F 1980 X-ray standing waves at crystal surfaces *Phys. Rev. Lett.* **44** 1680 https://doi.org/10.1103/PhysRevLett.44.1680

[44] Golovchenko J A, Patel J R, Kaplan D R, Cowan P L and Bedzyk M J 1982 Solution to the surface registration problem using x-ray standing waves *Phys. Rev. Lett.* **49** 560 https://doi.org/10.1103/PhysRevLett.49.560

[45] Bedzyk M J, Bommarito G M and Schildkraut J S 1989 X-ray standing waves at a reflecting mirror surface *Phys. Rev. Lett.* **62** 1376 https://doi.org/10.1103/PhysRevLett.62.1376

[46] Wang J, Bedzyk M J, Penner T L and Caffrey M 1991 Structural studies of membranes and surface layers up to 1,000 Å thick using x-ray standing waves *Nature* **354** 377 https://doi.org/10.1038/354377a0

[47] Bedzyk M J, Bilderback D H, Bommarito G M, Caffrey M and Schildkraut J S 1988 X-ray standing waves: a molecular yardstick for biological membranes *Science* **241** 1788 https://doi.org/10.1126/science.3175619

[48] Tegze M and Faigel G 1996 X-ray holography with atomic resolution *Nature* **380** 49–51 https://doi.org/10.1038/380049a0

[49] Takahashi T, Sumitani K and Kusano S 2001 Holographic imaging of surface atoms using surface x-ray diffraction *Surf. Sci.* **493** 36–41 https://doi.org/10.1016/S0039-6028(01)01186-4

[50] Okasinski J S, Kim C-Y, Walko D A and Bedzyk M J 2004 X-ray standing wave imaging of the 1/3 monolayer Sn/Ge(111) surface *Phys. Rev.* B **69** 041401(R) https://doi.org/10.1103/PhysRevB.69.041401

[51] Escuadro A A, Goodner D M, Okasinski J S and Bedzyk M J 2004 X-ray standing wave analysis of the Sn/Si(111)–$\sqrt{3}\times\sqrt{3}$ surface *Phys. Rev.* B **70** 235416 https://doi.org/10.1103/PhysRevB.70.235416

[52] Zegenhagen J 2019 X-ray standing waves technique: Fourier imaging active sites *Jpn. J. Appl. Phys.* **58** 110502 https://doi.org/10.7567/1347-4065/ab4dec

[53] Zhao W and Sakurai K 2019 X-ray standing wave technique with spatial resolution: in-plane characterization of surfaces and interfaces by full-field x-ray fluorescence imaging *Phys. Rev. Mater.* **3** 023802 https://doi.org/10.1103/PhysRevMaterials.3.023802

[54] Sakurai K and Iida A 1995 Analysis of specific interface of thin films by x-ray fluorescence using interference effect in total reflection *Adv. X-Ray Anal.* **39** 695–700 https://doi.org/10.1154/S0376030800023132

[55] Sakurai K 2004 Grazing-incidence x-ray spectrometry *X-Ray Spectrometry: Recent Technological Advances* ed K Tsuji, J Injuk and R Van Grieken (New York: Wiley) ch 5.1 https://doi.org/10.1002/0470020431

IOP Publishing

X-ray Color Imaging

Static and dynamic x-ray fluorescence for chemical element identification

Kenji Sakurai and Wenyang Zhao

Chapter 5

New opportunities in x-ray color imaging

5.1 Expanding vision

It is important to know that any scientific measurement provides information about only a limited part of the whole body. Figure 5.1 is a famous picture, 'The Blind Men and the Elephant' [1], which shows a group of blind men who have never seen an elephant and who try to argue about what the elephant is like by touching different parts of the elephant based on their own perceptions. One might touch the trunk and conclude that an elephant is like a snake, while another who touches the leg might think it is like a tree trunk. Each blind person perceives only a small aspect of the whole, so their individual perspectives are incomplete and often contradictory. This story has helpful implications for modern science, especially when studying complex problems.

Scientific research is divided into many small, specialized fields for deep understanding of each detail. No doubt, specialization is a necessary step for science. On the other hand, each field of research has been established as a convenient way of advancing science efficiently. Therefore, these approaches must be reformed according to their stage of development. Looking at materials science, which has been in existence for more than half a century, one might feel that research in this field is too divided by applications and types of materials. Although the division itself is quite natural and specialization has some advantages, it may not be very good for scientists to limit their thinking and activities to this narrow scope. Overspecialization tends to lead to the adverse effects of group blindness. As a result, it is easy to overlook important phenomena and some natural laws that lie just outside the self-imposed boundaries of specialization. In the late 1980s, high-T_c superconductors were discovered by Johannes Bednorz and Karl Alexander Müller [2], condensed matter physicists with a different perspective from that of conventional metallic superconductor specialists. Bednorz and Müller were awarded the 1987 Nobel Prize in Physics for their important breakthrough in the discovery of superconductivity in ceramic materials. Similar problems can be found in many scientific fields.

doi:10.1088/978-0-7503-3215-6ch5

Figure 5.1. 'The Blind Men and the Elephant,' by Martha Adelaide Holton and Charles Madison Curry [1]. Each blind man forms an idea by touching a different part of the elephant, and all of them are partly true. To understand and explain the elephant, the integration of all such pieces is important. This File:Blind men and elephant.png has been obtained by the authors from the Wikimedia website, where it is stated to have been released into the public domain. It is included within this book on that basis.

In analytical spectroscopy, communities of scientists are established for each different wavelength, from infrared to high-energy x-rays.

Therefore, in scientific practice, it is always important to understand the limitations of overly specialized perspectives and the importance of integrating different views. The importance of developing and improving scientific instruments can be understood by seeing them as a tool that supports a broad understanding of nature and reduces the blind spots as much as possible.

Because of its non-destructive nature, x-ray color imaging can help broaden our view of nature by cooperating and connecting with the observations of other analytical methods. Multimodal imaging, i.e. integrated interpretations of the same field of view of the samples using images obtained by several techniques with different analytical information, sensitivity and spatial resolution, will be of importance in the near future. This will contribute to a better understanding of the properties and function of the materials and/or biological system. X-ray color imaging will play a central role in the analysis.

5.2 Multiscale vision

Atomic-scale microscopy techniques are sometimes criticized for seeing only trees and their leaves and branches, without providing knowledge about the whole forest. Such criticism seems biased, because such ultrahigh spatial resolution has been one of the most valuable key technologies for seeing and understanding nature at the atomic and molecular levels. In other words, not only electron microscopy and scanning probe microscopy but also other existing analytical methods, including x-ray analysis, require such ultrahigh spatial resolution. It is extremely important to see trees and their leaves and branches.

On the other hand, other scales are also important. The natural world, including artificial worlds, is inhomogeneous, and such inhomogeneity often affects the function or property of a system. The spatial resolution required for observation then depends on the spatial scale of the inhomogeneity. Some methods are suitable for observing tiny scales, others for macroscopic scales. Therefore, it is important to combine them and to add some new methods at the mesoscopic scale. It is also important to extend the scale of observation for each analytical method.

Figure 5.2 shows the current state of typical temporal and spatial resolutions and ranges in x-ray fluorescence (XRF) and XRF imaging. Conventional XRF analyzes

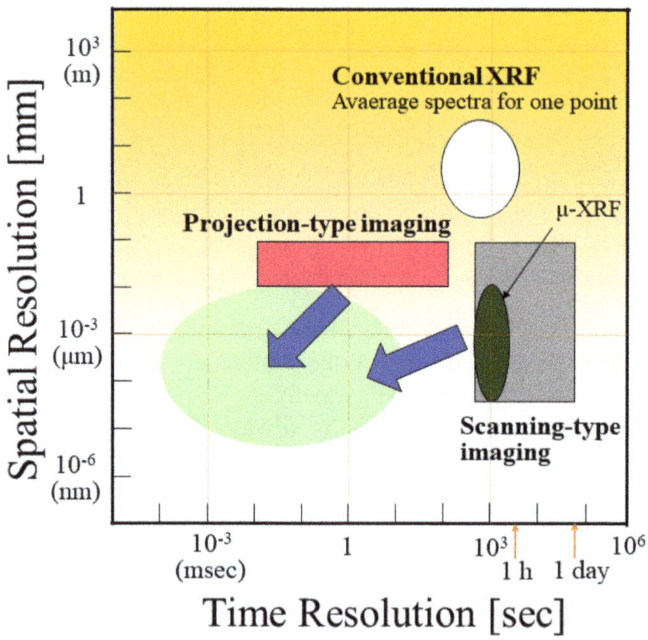

Figure 5.2. Spatial and temporal ranges in XRF spectroscopy and imaging (circa 2020). The scanning approach is superior in terms of spatial resolution, mainly because of the availability of nanofocusing. Since the projection approach does not require scanning, it is suitable for movie imaging applications. The time range is from 10 ms to a few minutes, but this will decrease to less than 1 ms. Another desirable major future breakthrough for projection-type imaging is spatial resolution better than 1 μm. Zoom capability is also required.

areal sizes from 0.1 mm to several centimeters using measurement times ranging from a few minutes (in the case of energy-dispersive XRF analysis) to about 1 h (in the case of wavelength-dispersive XRF analysis). Micro XRF can analyze much smaller areas with the help of focusing x-ray optics. In most laboratory systems, such areas are around 10–20 μm in size. Third-generation synchrotron beamlines can analyze areas in the range of 30 nm–10 μm. When such a small beam is used for scanning, this technique can produce an XRF image, i.e. scanning-type XRF imaging. Due to the elapsed time between the start and end points, the time resolution is limited in this case. On the other hand, the time resolution for projection-type XRF imaging is limited by the frame rate and some other technical factors in the 2D detector system. The time range for imaging is from 10 ms to a few minutes. On the other hand, the spatial resolution is still limited to about 10–20 μm. In the future, both scanning and projection methods will aim to improve their ability to image smaller objects with much shorter observation times. A continuous zoom capability would also be very attractive. X-ray color imaging with submicron spatial resolution (such as 50–100 nm) and a time resolution of about 0.1–1 ms could open a new field of research.

Most conventional 2D imaging detectors have used frame transfer to read the data accumulated after a preset time. Event-driven 2D detectors [3–6] could open up new possibilities in x-ray color imaging. This system transfers data only when photon events occur. The information to be sent to the computer is the time stamp, the pixel position(s) where the photon event occurred, and its charge. Simultaneous multielement imaging can easily be achieved by processing such event data in real time. Another attractive new possibility could be ultrafast x-ray color imaging in the picosecond-to-femtosecond range using a pump-and-probe scheme [7, 8].

5.3 Complex vision

There was a time when disciplines and fields of research were openly ranked. Victor Weisskopf, Director of CERN in 1965, once said that there is 'intensive' research in science (research that seeks fundamental truths, such as particle physics) and 'extensive' research (applied research in condensed matter physics, chemistry, biology, and other fields) [9]. Undoubtedly, fundamental physics is important. At that time, however, the complexity of nature was overlooked. Some physicists thought that everything could be explained just by applying knowledge obtained from fundamental research such as particle physics.

In 1972, P W Anderson responded with a paper entitled 'More is different,' arguing that there is a hierarchical structure in nature and that study X is not merely an application of study Y, but that there are new laws, new concepts, and new generalizations. New laws, new concepts, and new generalizations are needed, and this is where inspiration and creativity are required [10].

Condensed matter physics is not subordinate to particle physics or astrophysics. They are fundamentally different subjects of research. Chemistry is not a mere application of physics, and biology is not a mere application of chemistry. There was a time when what is taken for granted today was not always so. No matter how advanced science and technology may seem to have become, the unknown is still far greater than what is believed to be known. The names and classifications of the disciplines and fields of study were only tentatively established for the sake of convenience at that stage of development.

Today, a little more than 50 years after Anderson's paper, there are clearly more opportunities to study complex and nonlinear systems. In such systems, the tools of research need to be reinvented, as they are difficult to derive from the conventional sum of knowledge and information. When we consider the elucidation of various chemical pattern formations and chaotic phenomena in reaction–diffusion systems and their application as sensors, there are obvious limitations in many conventional analytical methods that assume that the sample is invariant. Static crystal structures and average chemical compositions are necessary information to carry out research, but it is not enough to know them; the core of the study of the system needs to be analyzed separately.

X-ray color imaging, which enables snapshot and movie imaging of chemical elements, is a promising technique that is suitable for complex systems. In the future, this technique will be further upgraded so that it will be fit to solve complex problems.

5.4 The use of artificial intelligence

Advances in imaging techniques, especially in time resolution, inevitably increase the size of the data collected by orders of magnitude. In x-ray color imaging using charge-coupled device (CCD) or complementary metal-oxide-semiconductor (CMOS) sensors, it is necessary to acquire a relatively large number of images due to the need for single-photon-counting conditions. Even at one point in time, it is necessary to process such data to obtain multielement images simultaneously. If the movie in question is fast, the amount of raw image data to be processed will be large. This is not a problem if a fast computer is used; the real problem is the detailed quantitative interpretation of the multielement images finally obtained.

Artificial intelligence (AI) emerged in the 1950s, around the same time as the advent of computers. The 1980s saw the second breakthrough of AI in the forms of expert systems, which took advantage of the engineering workstations of the time, as well as knowledge-based systems. After 1993, AI in the form of machine learning became promising due to the rapid development of computing power and the development of various theories of neural networks. Since 2011, big data and deep learning have dramatically improved the capabilities of AI [11, 12], especially in the fields of imaging and social networking, and its power and usefulness have been recognized in many fields. This third breakthrough is still ongoing.

On the other hand, in the field of x-ray imaging, the automation of the medical diagnosis of chest x-rays was slow for many years. A change happened after 2019 [13–15]. As AI has already contributed to efficient and successful image analysis in other fields, x-ray radiography will soon be revolutionized. When x-ray color imaging becomes able to provide big data, the same transition will happen. Future XRF image analysis could be performed via AI-driven high-speed automated analysis. Deep learning has already begun to penetrate the field of x-ray and neutron scattering and spectroscopic data [16]. The immediate goal is to automatically interpret and make some sort of decision on large amounts of data acquired almost unattended and without the assistance of knowledgeable experts.

An important future goal is not only to handle x-rays and other data but also to automatically search for new functional materials in chemistry and materials engineering. This has been proposed for several decades, but each time, it has fizzled out; now, it seems that this is the right time. Once the properties and functions are defined as targets, the latest AI proposes the candidate sets of the best materials with a prioritized list of recipes for their synthesis, using databases [17, 18]. The system is now being combined with robotics, leading to automated, unmanned chemical synthesis. It is now called self-driving laboratories [19–24]. On the other hand, new functional materials are still discovered by specialists every year, even by chance. The mindset of researchers who are not necessarily tied to the AI system based on large databases may remain dominant for some time to come. However, the future replacement of knowledgeable experts by AI systems is generally a not bad thing, because scientists will be able to afford to explore more new fields. X-ray color imaging will be used in such new scientific work.

5.5 Recommendations for the reader

X-ray color imaging is an extension of x-ray fluorescence analysis. It provides snapshot and movie images in addition to the usual analysis of average chemical composition. Since x-ray color imagers can also be used as conventional energy-dispersive x-ray fluorescence spectrometers, it can be said that they have backward compatibility.

X-ray color imaging is a new technique, but it is not difficult for anyone to use anywhere. If the reader already has an x-ray tube, an x-ray fluorescence spectrometer, or an x-ray diffractometer, this means that the reader can do x-ray color imaging. Even if it is not always easy to get access to sophisticated 2D energy-dispersive x-ray detectors, CCD and CMOS cameras can be good alternatives. In short, the reader can try x-ray color imaging today! Hopefully, this book will be useful in exploring the colorful natural world in terms of chemical elements.

References

[1] Holton M A and Curry C M 1914 *Holton–Curry Readers: The Fourth Reader* (Chicago, IL: Rand McNally and Co.) 108 http://archive.org/details/holtoncurryread01currgoog

[2] Bednorz J G and Müller K A 1986 Possible high T_c superconductivity in the Ba–La–Cu–O system *Z. Phys.* **B64** 189–93 https://doi.org/10.1007/BF01303701

[3] Mahowald M A and Mead C 1991 The silicon retina *Sci. Am.* **264** 76–83 https://doi.org/10.1038/scientificamerican0591-76

[4] Bailey G W *et al* 1997 Event-driven 2D detector for digital electron imaging *Microsc. Microanal.* **3** 1089–90 https://doi.org/10.1017/S1431927600012332

[5] Jannis D, Hofer C, Gao C, Xie X, Béché A, Pennycook T J and Verbeeck J 2022 Event driven 4D STEM acquisition with a Timepix3 detector: microsecond dwell time and faster scans for high precision and low dose applications *Ultramicroscopy* **233** 113423 https://doi.org/10.1016/j.ultramic.2021.113423

[6] Levin B D A, Spilman M and Bammes B 2023 Event-based direct detectors and their applications in electron microscopy *Wiley Microsc. Anal.* **64** 15–7 https://analyticalscience.wiley.com/content/article-do/event-based-direct-detectors-and-their-applications-electron-microscopy

[7] Kraus P M, Zürch M, Cushing S K, Neumark D M and Leone S R 2018 The ultrafast x-ray spectroscopic revolution in chemical dynamics *Nat. Rev. Chem.* **2** 82–94 https://doi.org/10.1038/s41570-018-0008-8

[8] Vassholz M *et al* 2021 Pump–probe x-ray holographic imaging of laser-induced cavitation bubbles with femtosecond FEL pulses *Nat. Commun.* **12** 3468 https://doi.org/10.1038/s41467-021-23664-1

[9] Weisskopf V 1965 In defense of high energy physics *Nature of Matter – Purposes of High-Energy Physics (BNL 888 (T-360))* ed L C L Yuan (Upton, NY: Brookhaven National Laboratory) pp 24–7 https://nemenmanlab.org/%7Eilya/images/a/a4/Weisskopf-1965.pdf

[10] Anderson P W 1972 More is different: broken symmetry and the nature of the hierarchical structure of science *Science* **177** 393–6 https://doi.org/10.1126/science.177.4047.393

[11] Hinton G E 2007 Learning multiple layers of representation *Trends Cogn. Sci.* **11** 428–34 https://doi.org/10.1016/j.tics.2007.09.004

[12] LeCun Y, Bengio Y and Hinton G E 2015 Deep learning *Nature* **521** 436–44 https://doi.org/10.1038/nature14539 https://nature.com/articles/nature14539

[13] Reardo S 2019 Rise of robot radiologists *Nature* **576** S54–8 https://doi.org/10.1038/d41586-019-03847-z

[14] Pisano E D 2020 AI shows promise for breast cancer screening *Nature* **577** 35–6 https://doi.org/10.1038/d41586-019-03822-8

[15] Savage N 2020 How AI is improving cancer diagnostics *Nature* **579** S14–6 https://doi.org/10.1038/d41586-020-00847-2

[16] Chen Z *et al* 2021 Machine learning on neutron and x-ray scattering and spectroscopies *Chem. Phys. Rev.* **2** 031301 https://doi.org/10.1063/5.0049111

[17] https://next-gen.materialsproject.org/materials/gnome

[18] https://github.com/google-deepmind/materials_discovery

[19] Szymanski N J, Nevatia P, Bartel C J, Zeng Y and Ceder G 2023 Autonomous and dynamic precursor selection for solid-state materials synthesis *Nat. Commun.* **14** 6956 https://doi.org/10.1038/s41467-023-42329-9

[20] Merchant A, Batzner S, Schoenholz S S, Aykol M, Cheon G and Cubuk E D 2023 Scaling deep learning for materials discovery *Nature* **624** 80–5 https://doi.org/10.1038/s41586-023-06735-9

[21] Szymanski N J *et al* 2023 An autonomous laboratory for the accelerated synthesis of novel materials *Nature* **624** 86–91 https://doi.org/10.1038/s41586-023-06734-w

[22] Bennett J A, Orouji N, Khan M, Sadeghi S, Rodgers J and Abolhasani M 2024 Autonomous Reaction Pareto-front mapping with a self-driving catalysis laboratory *Nat. Chem. Engin.* **1** 240–50 https://www.nature.com/articles/s44286-024-00033-5 https://doi.org/10.1038/s44286-024-00033-5

[23] Volk A A and Abolhasani M 2024 Performance metrics to unleash the power of self-driving labs in chemistry and materials science. Performance metrics to unleash the power of self-driving labs in chemistry and materials science *Nat. Commun.* **15** 1378 https://doi.org/10.1038/s41467-024-45569-5

[24] Tom G *et al* 2024 Self-driving laboratories for chemistry and materials science ChemRxiv https://doi.org/10.26434/chemrxiv-2024-rj946

Appendix A

The x-ray periodic table

Table A.1 shows the energies of the x-ray absorption edges (K, L_1, L_2, L_3 and M_5) and x-ray fluorescence peaks ($K\alpha_1$, $K\beta_1$, $L\alpha_1$, $L\beta_1$, $L\beta_2$, $L\beta_3$, $L\beta_4$, $L\gamma_1$, $M\alpha$, and $M\beta$) for all elements in the form of the periodic table [1]. The reader can download the same table as a digital image at several different resolutions which are suitable for use as a poster [2].

Table A.1. X-ray absorption and emission energies of the elements.

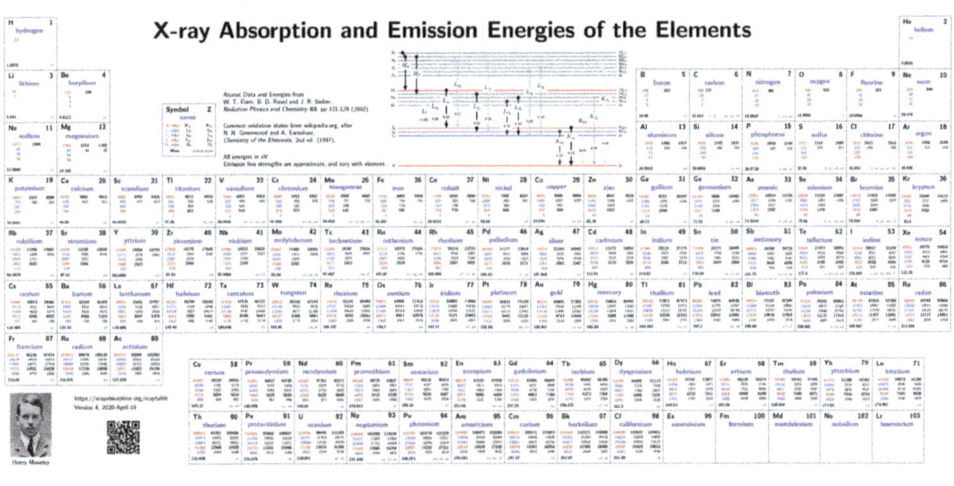

References

[1] Elam W T, Ravel B D and Sieber J R 2002 A new atomic database for X-ray spectroscopic calculations *Radiat. Phys. Chem.* **63** 121–128 https://doi.org/10.1016/s0969-806x(01)00227-4

[2] 2020 IXAS: The International X-ray Absorption Society. The Xray Periodic Table https://xrayabsorption.org/xraytable/

doi:10.1088/978-0-7503-3215-6ch6

IOP Publishing

X-ray Color Imaging
Static and dynamic x-ray fluorescence for chemical element identification
Kenji Sakurai and Wenyang Zhao

Appendix B

X-ray radiation safety

B.1 The biological effects of ionizing radiation

X-rays are photons that have enough energy to ionize atoms and molecules. Undoubtedly, x-rays are a type of ionizing radiation that may cause some biological health effects. Ionization in biological tissues can cause changes at the molecular level, including the genetic information clusters of cells, i.e. the DNA in the cell nucleus. Depending on the dose level, this can lead to short-term and/or long-term health effects in the form of radiation damage to organs and tissues of the body. The International Commission on Radiological Protection (ICRP) publishes recommendations for preventing accidents in radiation work. The latest recommendation was published in 2007 [1], and previous recommendations published in 1977 [2] and 1990 [3] provide useful and important information. Some publications on general principles for worker protection are also helpful [4]. Based on the ICRP recommendations, each nation regulates the use of ionizing radiation by law.

In general, at extremely high doses, which are not normally expected in x-ray experiments, severe damage to organs and tissues can occur. This can sometimes lead to loss of function and cell death. When the doses are quite low, such reactions are not seen. However, some damage at the genomic level can increase the risk of future cancer. These two types of effects are usually referred to as deterministic and stochastic effects, respectively. Table B.1 summarizes the dose limits [1].

B.2 Practical tips for radiation safety in real x-ray experiments

X-ray equipment must always be shielded. When the shutter is open and x-rays are running, all adjustments and measurements must be made by remote control. X-ray experiments are almost always made safe by removing access to the x-ray pathways. In other words, an interlock system is extremely important for safety.

Many x-ray sources use cooling water. Depending on the humidity and temperature of the laboratory, condensation can form on the surface of the x-ray tube. This can cause the surface of the shutter plate, which is usually made of lead, to rust. When the

doi:10.1088/978-0-7503-3215-6ch7

Table B.1. Recommended dose limits in planned exposure situations [1].

Type of limit	Occupational	Public
Effective dose	20 mSv per year, averaged over defined periods of 5 years	1 mSv in a year
Annual equivalent dose in:		
Lens of the eye	150 mSv	15 mSv
Skin	500 mSv	50 mSv
Hands and feet	500 mSv	—

surface becomes rough, the shutter may not close properly in response to the remote electronic signal. To maintain the reliability of the interlock system, it is highly recommended that the shutter system be checked, especially on older machines.

Unlike radiation damage caused by radioactive material, an x-ray exposure accident may affect only part of the body, such as the fingers, hands, or face, rather than the entire body. This type of accident is most likely to occur when manual adjustments of optical components and the sample are attempted in the presence of x-rays—in other words, by intentionally disabling the interlock.

The risk of damage depends on the beam size and divergence and the x-ray energy. When a monochromator is used in an experiment, the beam size and its divergence are limited, so the risk is much lower than with white (continuous spectrum) x-rays. If we consider copper and molybdenum x-ray tubes, for example, experiments with a molybdenum tube require more caution due to the difference in energy. When the tube voltage is increased, the primary x-rays include higher-energy x-rays, resulting in an increase in the risk.

The intensity of the primary x-rays is much higher than that of the x-ray fluorescence emitted by the sample. From the point of view of radiation exposure risk, primary x-rays are more dangerous, even if a low-power x-ray source is used for the experiment.

References

[1] International Commission on Radiological Protection (ICRP) 2007 Therecommendations of the International Commission on Radiological Protection *ICRP Publication 103. Ann. ICRP 37* No. 2-4 (2007). https://icrp.org/publication.asp?id=ICRP%20Publication%20103

[2] International Commission on Radiological Protection (ICRP) 1977 Recommendations of the ICRP *ICRP Publication 26. Ann. ICRP 1* No. 3. https://icrp.org/publication.asp?id=ICRP%20Publication%2026

[3] International Commission on Radiological Protection (ICRP) 1991 1990 Recommendations of the International Commission on Radiological Protection *ICRP Publication 60. Ann. ICRP 21* No. 1-3. https://icrp.org/publication.asp?id=ICRP%20Publication%2060

[4] International Commission on Radiological Protection (ICRP) 1997 General principles for the radiation protection of workers *ICRP Publication 75. Ann. ICRP 27* No. 1. https://icrp.org/publication.asp?id=ICRP%20Publication%2075

www.ingramcontent.com/pod-product-compliance
Ingram Content Group UK Ltd.
Pitfield, Milton Keynes, MK11 3LW, UK
UKHW051851100926
8556IPUK00005B/566